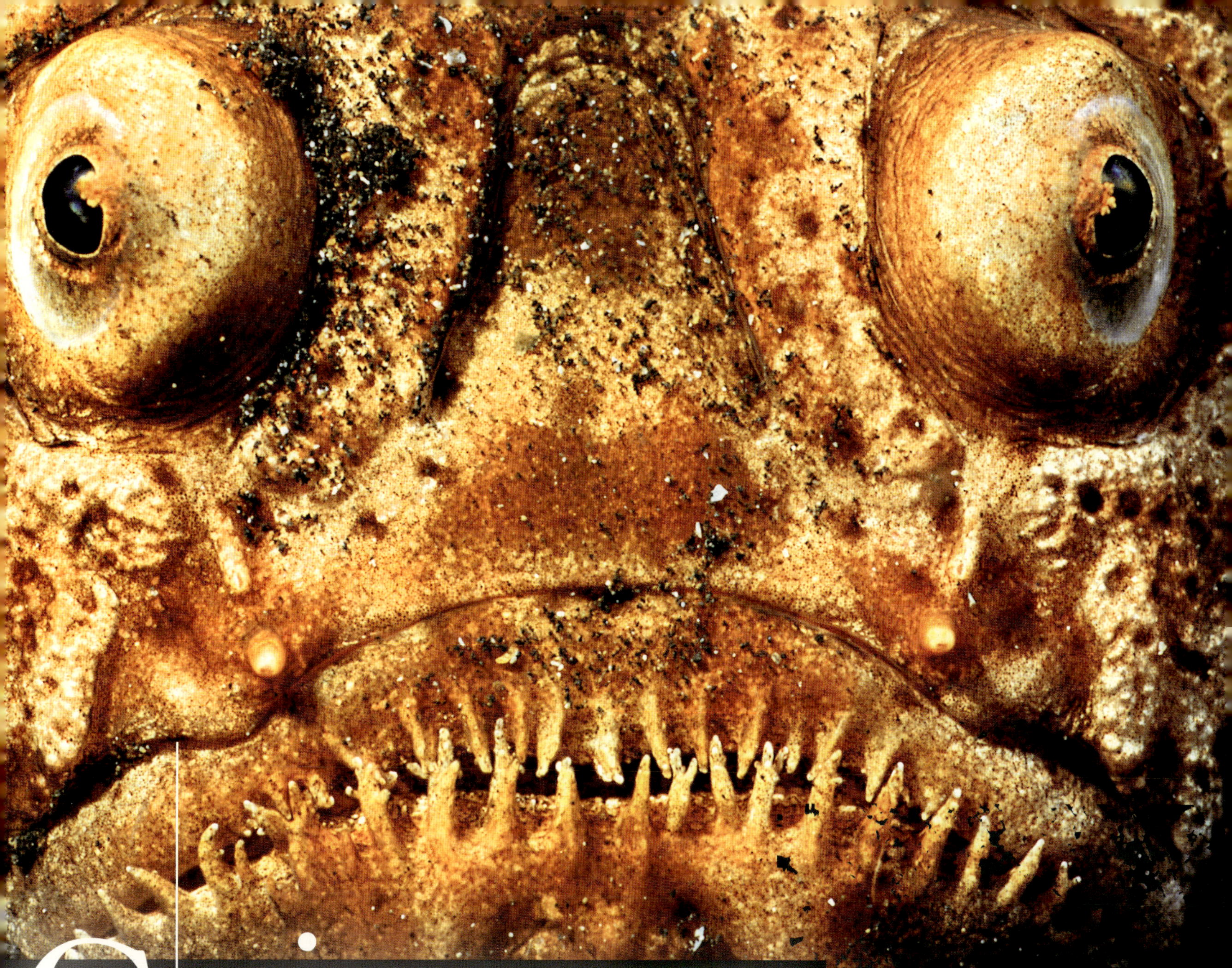

Critters

mysterious
twilight zone

monique walker

Für meine Mutter Frida Walker

To my mother Frida Walker

A comment on this book

Mankind is already exploring the stars. The seven seas, however, remain full of secrets from us.
Diving into the underwater world we hover astronaut-like, suspended in an unfamiliar element, marvelling at the wonders of this mysterious habitat. There is a stunning variety of species among marine creatures. Many of them are known to the wider public through the medium of films and literature. However, the more bizarre creatures have hardly ever been revealed in this way. This book is my best attempt to display some of the most weird and bizarre underwater "critters" I have encountered during my diving experience.

After publishing my first book, which laid the main stress on the topic of marine biology, I have now undertaken the task of compiling this book to present subjects which constitute my secret passion – the "critters" of the tropical seas. These less common creatures have an incredibly mysterious and mystic aura. Finding and observing such often gruesome, though marvellous creatures, is an unforgettable experience.

I want to take this opportunity to express my heartfelt thanks to the persons involved in the production of this book, and who have supported this project by other means.

My thanks to Dr. John E. Randall, the author of the English preface, and to Dr. Harry Erhardt who wrote the German preface and made biological classifications, to Max Gibbs for checking the translations and classifications.

Furthermore I want to give many thanks to my understanding companion and diving partner, Dr. Wolfgang K. Pfeiffer, Herta Thiel, Michael Bauer, Diana Pauser, Ute Rockenfeller, Dr. Günter & Renate Therhag, as well as Captain Tore Thorsen (Norway), Captain Lance Higgs (Micronesia), and Divemaster Larry Smith.

Monique Walker

Frankfurt, Spring 2000

zu diesem Buch

Die Menschheit erkundet bereits die Sterne, aber noch sind die Weltmeere für uns voller Überraschungen. Unter Wasser schweben wir gleich den Astronauten durch das uns fremde Element und besuchen staunend diesen geheimnisvollen Lebensraum. Die Artenvielfalt der Meeresbewohner ist faszinierend. Zwar sind viele Meereslebewesen heutzutage durch Film und Literatur recht gut bekannt, jedoch nicht die ungewöhnlicheren Tiere. Daher soll dieser Bildband vorwiegend seltene und skurrile Unterwasserbegegnungen dokumentieren. Nach meinem ersten Buch mit Schwerpunkt Meeresbiologie habe ich es mir zur Aufgabe gemacht, ein Werk über meine heimliche Passion – die „Critters" – zu verfassen. Diese nicht alltäglichen Meereslebewesen verfügen über eine unglaublich geheimnisvolle und mystische Ausstrahlungskraft. „Critters", diese gruseligschönen Kreaturen zu entdecken und zu beobachten, ist ein einzigartiges Erlebnis.

An dieser Stelle möchte ich mich bei allen bedanken, die zur Entstehung dieses Werkes beigetragen oder dieses in anderer Weise unterstützt haben:

Mein besonderer Dank gilt Dr. John E. Randall für das englische Vorwort, Dr. Harry Erhardt für das deutsche Vorwort und die biologischen Bestimmungen sowie Max Gibbs für die Bearbeitung der Übersetzungen und die fachkundigen Bestimmungen.

Ebenso geht mein Dank an meinen verständnisvollen Lebensgefährten und Tauchpartner Dr. Wolfgang K. Pfeiffer, an Herta Thiel, Michael Bauer, Diana Pauser, Ute Rockenfeller und Renate & Dr. Günter Therhag sowie an Captain Tore Thorsen (Norwegen), Captain Lance Higgs (Mikronesien) und Divemaster Larry Smith.

Monique Walker

Frankfurt, Frühjahr 2000

foreword
john e. Randall, phD

Most underwater photographers who dive in tropic seas prefer to aim their cameras at the colorful marine life in the crystalline water of coral reefs, or they may concentrate on getting photographs of sharks, manta rays, or other large fishes such as groupers and jacks. A growing number of photographers have discovered that there are some remarkable marine animals that may be found in other habitats, often where the sea is not so clear, such as seagrass beds, sandy bottom, or even mud bottoms of mangrove swamps. Some of the fishes and invertebrates in these areas are colorful, but most are camouflaged and difficult to see. The real challenge is just to find them. Monique Walker is very special in her ability to not only spot these cryptic animals but in doing a magnificent job of capturing them on film. I am pleased to see that she has produced this volume of her beautiful images of marine life for all to enjoy. I note that most of the pictures are taken in Indonesia. Along with the Philippines and Papua New Guinea, this East Indian region has the richest marine fauna and flora in the world. For fishes alone, it is estimated that there are 3000 species in the East Indies. Among the least known are those from the habitats other than coral reefs. Marine biologists like myself have also concentrated on the coral reef fishes, but we need to spend more time investigating the life of the other marine habitats.

I met Monique in 1995 when diving at Laha, Ambon in the Molucca Islands. Our first night dive resulted in an experience which I know she has not forgotten because of the laughter that it produced. I speared a large snake eel, and after pulling it from the silty sand bottom, I tried to get it off the spear into a net bag. It escaped my grasp, and instead of returning to the bottom, it went straight for me and bit me in the crotch. Fortunately I was wearing wet suit pants, so its teeth seized only foam rubber. The eel is now a museum specimen.

The brief stay at Laha was followed by a memorable dive cruise on Tore Thorsen's "Cehili". Off the east coast of Seram one of the divers caught a bright green filefish and brought it to the boat. I wanted to photograph it, as did Monique and others, but it was too late in the day. We kept it alive overnight, and I liberated it the next morning in its seagrass habitat where its green color matched that of the plants perfectly. Four underwater photographers surrounded the fish and repeatedly flashed it with their strobes. Later I learned that Monique had won a first prize in an underwater film contest with her photo of the fish.

As the reader will note, all of her photographs herein are prize-winners.

John E. Randall, PhD

Bishop Museum, Honolulu, USA
Spring 2000

vorwort
john e. Randall, phD

Die Mehrzahl der Unterwasser-Fotografen richten ihre Kameras auf die farbenprächtige Unterwasserwelt im klaren Wasser der Korallenriffe, oder sie fotografieren Haie, Mantas, oder andere große Fische wie Barsche oder Makrelen. Immer mehr Fotografen haben aber entdeckt, daß auch andere Lebensräume wie trübes Wasser, Seegraswiesen, Sandböden oder sogar Schlammböden von Magrovensümpfen von bemerkenswerten Meerestieren bewohnt werden. Einige dieser Fische und Wirbellosen sind farbenprächtig, die meisten jedoch getarnt und schwierig zu sehen. Sie zu finden ist eine wahre Herausforderung. Monique verfügt über die spezielle Fähigkeit, diese versteckt lebenden Tiere zu finden und diese meisterhaft zu fotografieren. Es freut mich sehr, daß sie diesen Bildband mit ihren schönen Bildern von Meereslebewesen zur Freude aller produziert hat. Die meisten Aufnahmen sind in Indonesien gemacht worden. Mit den Philippinen und Papua New Guinea beherbergen diese ostindischen Regionen die artenreichtste Fauna und Flora der Welt. Nur alleine die Fischarten werden im ostindischen Raum auf über 3000 geschätzt. Die unbekanntesten sind die, die in anderen Lebensräumen als im Korallenriff zu finden sind. Meeresbiologen, wie ich selbst, haben sich auf Fische im Korallenriff konzentriert, aber wir müssen zukünftig verstärkt die Meereslebewesen in anderen Lebensräumen erforschen.

Ich habe Monique 1995 in Laha, Ambon in den Mollukken kennengelernt. Unseren ersten Nachttauchgang wird sie durch unser gemeinsames Erlebnis und das daraus resultierende Gelächter bestimmt nie vergessen. Nachdem ich einen großen Schlangenaal gespeert hatte, zog ich ihn aus dem schlammigen Boden. Ich versuchte ihn vom Speer zu nehmen und in meiner Netztasche zu verstauen. Dieser entwischte jedoch meinem Griff und statt auf den Boden zurückzukehren, ging er direkt auf mich los und biss mich zwischen die Beine. Glücklicherweise trug ich Neoprenhosen und seine Zähne konnten den Schaumstoff nicht durchdringen. Der Aal ist heute im Museum zu bewundern.

Unserem kurzen Aufenthalt in Laha folgte eine erinnerungswürdige Tauchkreuzfahrt mit Tore Thorsen's „Cehili". Vor der Ostküste Seram's hatte ein Taucher einen grünen Feilenfisch gefangen und ihn zu wissenschaftlichen Zwecken an Bord gebracht. Ich wollte ihn fotografieren und Monique und die anderen auch, aber es war bereits zu spät. So blieb er über Nacht an Bord und ich brachte ihn am nächsten Morgen zurück in seinen Lebensraum, wo er sich mit seiner grünen Färbung perfekt dem Seegras anpasste. Vier Fotografen verteilten sich um den Fisch und ein Blitzgewitter begann. Später hörte ich, daß Monique mit ihrem Bild von diesem Fisch einen Wettbewerb gewonnen hatte. Wie Sie feststellen können, sind alle ihre Aufnahmen in diesem Bildband Preisgewinner.

John E. Randall, PhD

Bishop Museum, Honolulu, USA
Frühjahr 2000

foreword vorwort

Dr. Harry Erhardt

The oceans on this planet promise to provide the last adventures for exploring the world's fauna. They are jealously guarding many surprises for sure, and not only for marine biologists. This refers to the bizarre deep-sea marine fauna and the underwater caves which are hardly known and reported about in scientific literature to date. While wealthy oil and offshore concerns are spending millions on the construction and use of sophisticated equipment to explore and exploit the deep sea, prospecting for gas, oil, and other valuable resources, Monique Walker is on excursions into the shallow water areas and reefs of tropical seas equipped only with comparatively modest means, in order to discover and photograph the so-called "critters". Many of these animals have an appearance just as bizarre as many of the invertebrates and fish from the ocean's greatest depths.

In American terminology the term "critters" is usually used to describe creatures which occupy an ecological niche in their habitat, being specifically adapted to the needs of their environment in order to survive. In particular they are animals which camouflage themselves by perfectly adapting to their surroundings, evolving colours and body structures which enables them to enjoy the ability to blend into their surroundings convincingly, avoiding the attentions of their natural enemies. With this perfect camouflage they lie in wait ready to ambush their prey as it chances by, unaware of the hidden predator.

I imagine that just the title of this book will grasp your attention and curiosity, causing you to pick it up and browse through its pages. You will certainly be amazed at the pictures you find here. Monique Walker has particularly selected and arranged these treasures from her extensive photo archives. The result is this fascinating collection of the "critters" she has painstakingly sought, discovered, and "shot" during thousands of hours underwater both by day and by night. I have accompanied Monique on some of her journeys through the fantastic underwater world. I am fascinated by her never ending enthusiasm for the animals and life within the reef, and I admire her eagerness to learn something of their biological features and pass this wisdom on to others.

During the many hours Monique has spent on the reef she has trained her eyes to specially find the many creatures which live largely hidden away from our less perceptive eyes. During every dive she makes a thorough search of the reef to find these "critters", always finding them sooner or later, even though they may measure just a few millimetres. Her camera is a tool she handles masterfully to take photographs of whatever she sees which commands her interest.

I feel sure this book will be very well received, with its broad appeal. To present rare shallow-water creatures is something exciting and special, which has not been offered on the book market in this way before. I wish Monique and all of us many more such beautiful books about the "critters" of the sea.

Dr. Harry Erhardt

Institute for Zoology Kassel, Germany
Spring 2000

Die Meere auf unserem Planeten versprechen noch eines der letzten großen Abenteuer in der Erforschung der Tierwelt zu werden, was nicht nur für die Meeresbiologen so manche Überraschung bereithalten wird. Man denke dabei an die skurrile Tierwelt der Tiefsee und die der unterseeischen Höhlen, von der auch in der wissenschaftlichen Literatur bis heute sehr wenig bekannt und nachzulesen ist.

Während finanzstarke Öl- und Offshorefirmen Milliarden in die Konstruktion und den Einsatz von modernstem Gerät zur Erforschung und Ausbeutung der Tiefsee ausgeben, um hier nach Gas, Öl und Erzlagerstätten zu suchen, ist Monique Walker mit außerordentlich bescheidenen Mitteln im Flachwasser tropischer Meere unterwegs, um mit ihrer Kamera ihre „Critters" aufzuspüren und sie abzulichten, Tiere, die teilweise ein ähnlich skurriles Aussehen haben wie so manche Wirbellose und Fische der Tiefsee.

Das Wort „Critters" ist im amerikanischen Sprachgebrauch nichts anderes als eine umschriebene Bezeichnung für Kreaturen, Lebewesen also, die versteckt und besonders angepaßt in ihrem Verbreitungsgebiet eine ökologische Nische besetzen, in der sie sich adaptieren, um zu überleben. Im Besonderen sind es Tiere, die sich anpassen, um nicht aufzufallen und dadurch vor ihren Freßfeinden sicher zu sein. Aus dieser perfekten Tarnung heraus betätigen sie sich selbst in der Regel als Überraschungs- und Lauerjäger, die ihre Beute aus dem Hinterhalt überwältigen.

Ich kann mir vorstellen, daß schon alleine der Buchtitel so viel Neugierde erweckt, daß man das Werk zur Hand nehmen und durchblättern muß. Erst dann kommt Staunen auf für das, was Monique Walker aus ihrem umfangreichen Bildarchiv ausgesucht und zusammengestellt hat. Es sind ihre „Critters", die sie in Tausenden von Stunden bei Tag und in der Nacht unter Wasser gesucht, gefunden und fotografiert hat.

Ich habe Monique auf einigen Reisen in die phantastische Welt unter Wasser begleitet und ich bin fasziniert von ihrer unerschöpflichen Begeisterung für die Tiere und das Leben im Riff und von ihrem Interesse, biologische Zusammenhänge zu erfahren, sie aufzunehmen und weiterzugeben.

Monique hat in den vielen Stunden, die sie im Riff verbrachte, ganz besonders ihre Augen dafür geschult, das wahrzunehmen, was mehr oder weniger im Verborgenen lebt. Bei jedem Tauchgang sucht sie sehr aufmerksam das ganze Riff nach ihren „Critters" ab und sie findet sie auch, egal, ob sie einige Zentimeter oder nur wenige Millimeter groß sind. Und ihr Werkzeug ist ihre Kamera, die sie meisterlich handhabt und mit der sie das fotografiert, was sie sieht und was sie interessiert.

Ich bin sicher, daß das vorliegende Buch Erfolg haben wird. Seltene Tiere des Flachwassers so schön und meisterlich in Szene gesetzt zu sehen, das ist schon etwas Aufregendes, etwas Besonderes, was auf dem Buchmarkt bisher fehlte.

Ich wünsche Monique und auch uns allen noch viele solcher schönen Bücher über die „Critters" im Meer.

Dr. Harry Erhardt

Zoologisches Institut Kassel, Deutschland
Frühjahr 2000

The "critters" — obscure hunting subjects

"Critters" are bizarre, unusual, and utterly rare sea creatures, whose appearance ranges from the weird and gruesome to even breath-taking beauty! Most amazing is the ability of many "critters" to adapt to their underwater environment with colour, structure, and form to such a degree that they become undetectable within their chosen surroundings. Photographic techniques can transform the more ordinary sea creatures into "critters" as well. For almost every observer "critters" can cast a spell of fascination, like an involuntary infection, promoting a desire to see ever more such natural wonders.

"critter-hunting" — the ultimate underwater sport

Among underwater photographers "critter-hunting" is gaining more and more dedicated followers, often travelling across the world to dive in unbeautiful areas in search of the weird and the wonderful. The term "Muck Diving" describes this activity very well. But "critters" are to be found in the most pristine coral reef environments too, of course. These days even non-photographers are descending with optically-corrected diving masks and magnifying glasses to find and marvel at the smallest of these animals.

The way my secret passion started

In 1994/95 I was onboard the diving cruiser, M.V. Cehili several times. Divemaster Larry Smith was the one to initially infect me with the "critters-virus" during that time! His sharp and experienced eyes never miss any "critter", no matter how tiny. Captain Tore Thorsen, the owner of the Cehili, had discovered some of my favourite diving grounds, which inspired me to write the subsequent report entitled "Ghost Train", and to create the awarded slide show, "Train Fantôme Sousmarin". Together with Dr. John E. Randall (known as "Jack" by his friends) the renowned ichthyologist and former curator of the Department of Ichthyology at the Bishop Museum, Honolulu, we were having a great and unforgettable time. Jack is the author of more than 500 scientific papers and several guide books on fishes. During his nearly 50 years of work he has described over 400 species of newly discovered reef fish. Even at his present age of 75 the "Fish Pope" is still regularly travelling away on diving expeditions, and is unceasingly busy. I will certainly never forget how he once asked me for help underwater. Unfortunately the discovered fish would not easily allow him to take a photo, but Jack did not give up. His efforts at 30 metres depth did not seem to trouble him at all, but I was exhausted, and found myself deeply impressed by his dedication to his scientific diving work.

During subsequent photographer's travels organised by Dive Expeditions I could enlarge my "critter" picture collection step-by-step. Invertebrates expert Dr. Harry Erhardt, and the aquarium and freshwater photographer, Max Gibbs, joined us regularly, enriching our interesting and adventurous diving expeditions. Apart from the marine biological highlights encountered, above all I appreciated the companionship and cheerfulness during these diving expeditions, and will always keep them fondly in my memory. The Oceans are so vast, surely holding so much more in store which will be thrilling to see, and concealing innumerable secrets waiting to be discovered.

Ghost Train

Beyond the beam of my underwater torch everything seemed so black and sinister. The closer I was diving back to the ship the more intensely and brightly the familiar rays of the ship's lighting penetrated the depths of the harbour basin. Before I had always felt regret when dive went by so fast. This time, however, I even felt relieved when getting back on the ship. First of all I had to take in all these new impressions. After my camera, mask and fins were taken by the crew, I silently climbed aboard the M.V. Cehili. Tore Thorsen hastened towards me, expecting my usual enthusiastic accounts, but this time I remained speechless. I went to my place on the dive deck, where I took off the rest of my diving equipment. Tore Thorsen sought an account from my diving partner, Divemaster Larry, but he did not seem to be very voluble either.

I was sitting there, stunned by the impressions of this night dive, and my thoughts were wandering back again into the depths. I was enchanted by the unique beauty of all the encounters I had experienced, and at the same time I was strangely exhilarated.
The more horrifying some of those underwater creatures appeared to me, the more impressed I got by facing them. Never before had I discovered so many bizarre fish and invertebrates during one single dive. Some reminded me of monsters with horrible grimaces. Ghost-like gigantic crabs, horrifying eels, toadfish camouflaged beyond recognition, magical ghost moray eels, frightening devilfish, and last but not least the big single-eyed mantis shrimp, besieged by small crabs as it hid in its burrow-lair. Particularly at night "critters" tend to evoke a strangely exhilarating yet beautiful shudder, and I had the feeling that this experience reminded me of something quite familiar. Yes, of course, I felt like having survived a descent into hell, viewing the spooky world of a ghost train.

The "critter" fever

Ever after this dive I have found myself definitely addicted towards the "critters", they have become my prime favourite subject for underwater photography.
Take the opportunity and dive down into a mysterious world, allowing yourself to become enchanted by such unusual underwater encounters.

Der „Critter“ — das obskure Objekt der Begierde

„Critters“ sind skurrile, außergewöhnliche und seltene Meereslebewesen. Sie haben ein gruseligschauriges bis wunderschönes Aussehen und können sich oft durch Farbe, Struktur oder Form bis zur Unerkennbarkeit an ihrem Aufenthaltsort in die Unterwasserwelt einfügen.
Je nach fotografischer Aufnahmetechnik können sich aber auch gewöhnlichere Meerestiere in „Critters“ verwandeln.
„Critters“ haben etwas Geheimnisvolles und Mystisches, das den Betrachter automatisch in seinen Bann zieht.

„Critter-Hunting“ — der ultimative Unterwasser-Sport

Unter Unterwasser-Fotografen erfreut sich das „Critter-Hunting“ immer größerer Beliebtheit. Auch Nicht-Fotografen tauchen immer häufiger mit optisch korrigierten Taucherbrillen und Lupen, damit ihnen auch der kleinste „Critter“ nicht entgeht.

Der Beginn einer heimlichen Leidenschaft

1994/95 war ich verschiedene Male mit dem Tauchkreuzfahrtschiff M.V. Cehili unterwegs.
Divemaster Larry Smith war damals derjenige, der mich endgültig mit dem „Critter-Virus“ infizierte. Seinem erfahrenen Auge entgeht kein „Critter“.
Captain Tore Thorsen, der Besitzer der M.V. Cehili, war seinerzeit auch der Entdecker einer meiner liebsten Tauchplätze, der mich zum Schreiben des anschließenden Berichtes „Geisterbahn“ und zur Idee des prämierten Diaporamas „Train Fantôme Sousmarin“ inspiriert hat.

Mit Dr. John E. Randall, Jack, wie er von seinen Kollegen und Freunden genannt wird, dem weltbekannten Ichthiologen und früheren Kurator der Abteilung Ichthiologie des Bishop Museums von Honolulu, verbrachten wir damals eine höchst interessante und unvergeßliche Zeit. Jack Randall ist Autor von über 500 wissenschaftlichen Abhandlungen und mehreren Fischbestimmungs-Büchern. Während seiner fast 50jährigen Tätigkeit beschrieb und benannte er über 400 neu entdeckte Riff-Fische. Obwohl inzwischen 75 Jahre alt, ist der „Fisch-Papst“ noch immer regelmäßig unterwegs und unermüdlich tätig.
Ich vergesse nie, wie er mich einmal bat, ihm unter Wasser zu helfen. Leider wollte sich der damals von ihm entdeckte Fisch nicht so einfach fotografieren lassen. Aber Jack gab nicht auf. Die Anstrengung in 30 m Tiefe schien ihm überhaupt nichts auszumachen. Ich dagegen war außer Atem und von seiner sportlichen Leistung tief beeindruckt.

Auf den vielen folgenden von Dive-Expeditions organisierten Fotografen-Reisen konnte ich meine „Critter-Sammlung“ nach und nach erweitern. Diese interessanten und erlebnisreichen Tauchreisen wurden oft vom Wirbellosen-Spezialisten Dr. Harry Erhardt und vom Aquarium- und Freiwasser-Fotografen Max Gibbs begleitet und bereichert. Abgesehen von den meeresbiologischen Highlights schätzte ich vor allem die Gesellschaft und die Heiterkeit dieser Tauch-Expeditionen, und ich werde diese Reisen immer in guter Erinnerung behalten. Allen zukünftigen Reiseplänen schaue ich mit großer Freude und Erwartung entgegen. Die Weltmeere sind so groß und bergen noch so viel Sehenswertes und unzählige Geheimnisse.

Geisterbahn

Außerhalb des Lichtkegels meiner Unterwasserlampe war alles schwarz und unheimlich. Je näher ich wieder an das Schiff zurücktauchte, desto intensiver und heller drang der vertraute Schein der Schiffsbeleuchtung in die Tiefe des Hafenbeckens. Sonst habe ich es immer bedauert, daß ein Tauchgang so schnell vorbeigeht. Aber diesmal war ich sogar froh, daß er ein Ende hatte. Ich mußte all die Eindrücke erst einmal verkraften. Nachdem mir Fotoapparat, Maske und Flossen abgenommen worden waren, stieg ich wortlos an Bord der M.V. Cehili. Tore Thorsen eilte mir entgegen und erwartete meine bereits gewohnten, begeisterten Erzählungen. Diesmal blieb ich jedoch schweigsam. Ich begab mich auf meinen Platz auf dem Tauchdeck, wo ich mich meiner restlichen Tauchausrüstung entledigte. Tore Thorsen versuchte sein Glück bei meinem Tauchpartner Divemaster Larry, aber auch der schien an diesem Abend nicht sehr gesprächig zu sein.

Von den Eindrücken dieses Nacht-Tauchgangs wie gelähmt, saß ich da, und meine Gedanken wanderten wieder zurück in die Tiefe. Ich war bezaubert von der einzigartigen Schönheit des Gesehenen und gleichzeitig erfüllt von einem Gefühl faszinierenden Gruselns. Je erschreckender mir einige Unterwasser-Lebewesen erschienen, desto tiefer war ich davon beeindruckt.
Noch nie zuvor hatte ich so viele skurrile Wirbellose und Fische während eines einzigen Tauchganges entdeckt. Einige sahen aus wie Monster mit schrecklichen Fratzen. Gespenstische, riesige Krabben, schreckenerregende Aale, bis zur Unkenntlichkeit getarnte Krötenfische, magisch anmutende Geistermuränen, furchterregende Teufelsfische, und nicht zu vergessen, der große, einäugige Heuschreckenkrebs in seiner von kleinen Krabben belagerten Wohnhöhle.
Besonders nachts vermitteln „Critters“ einen schrecklich-schönen Schauer, und ich hatte das Gefühl, daß mich dieses Erlebnis an etwas Vertrautes erinnert.
Ja, aber natürlich … ich fühlte mich wie nach einer Höllenfahrt durch die Gruselwelt einer Geisterbahn.

„Critteritis“

Seit diesem Tauchgang bin ich endgültig den „Critters“ verfallen, und sie sind zu meinen Lieblingsmotiven der Unterwasser-Fotografie geworden.
Tauchen Sie ab in eine geheimnisvolle Welt, und lassen auch Sie sich von nicht alltäglichen Unterwasser-Begegnungen verzaubern.

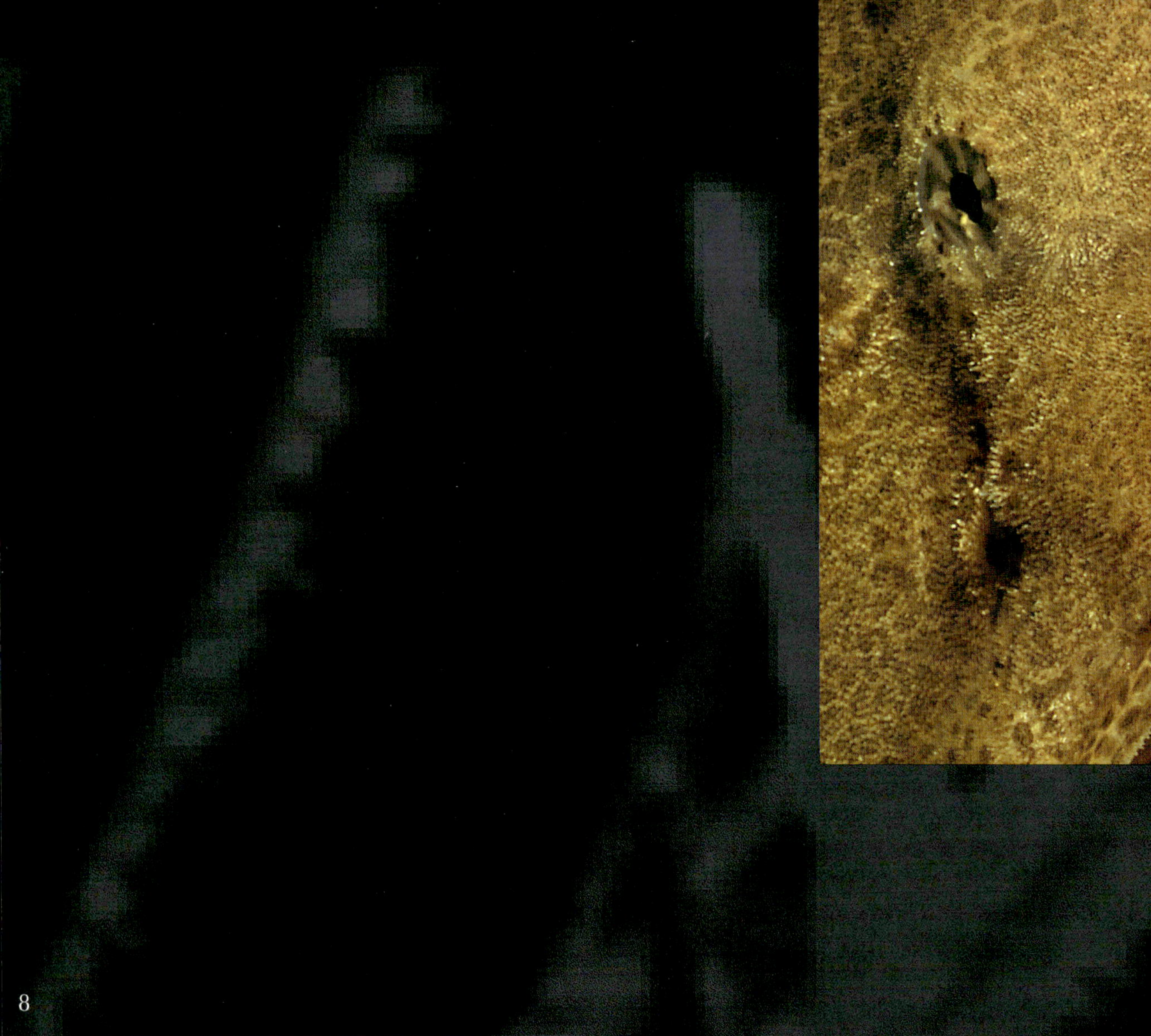

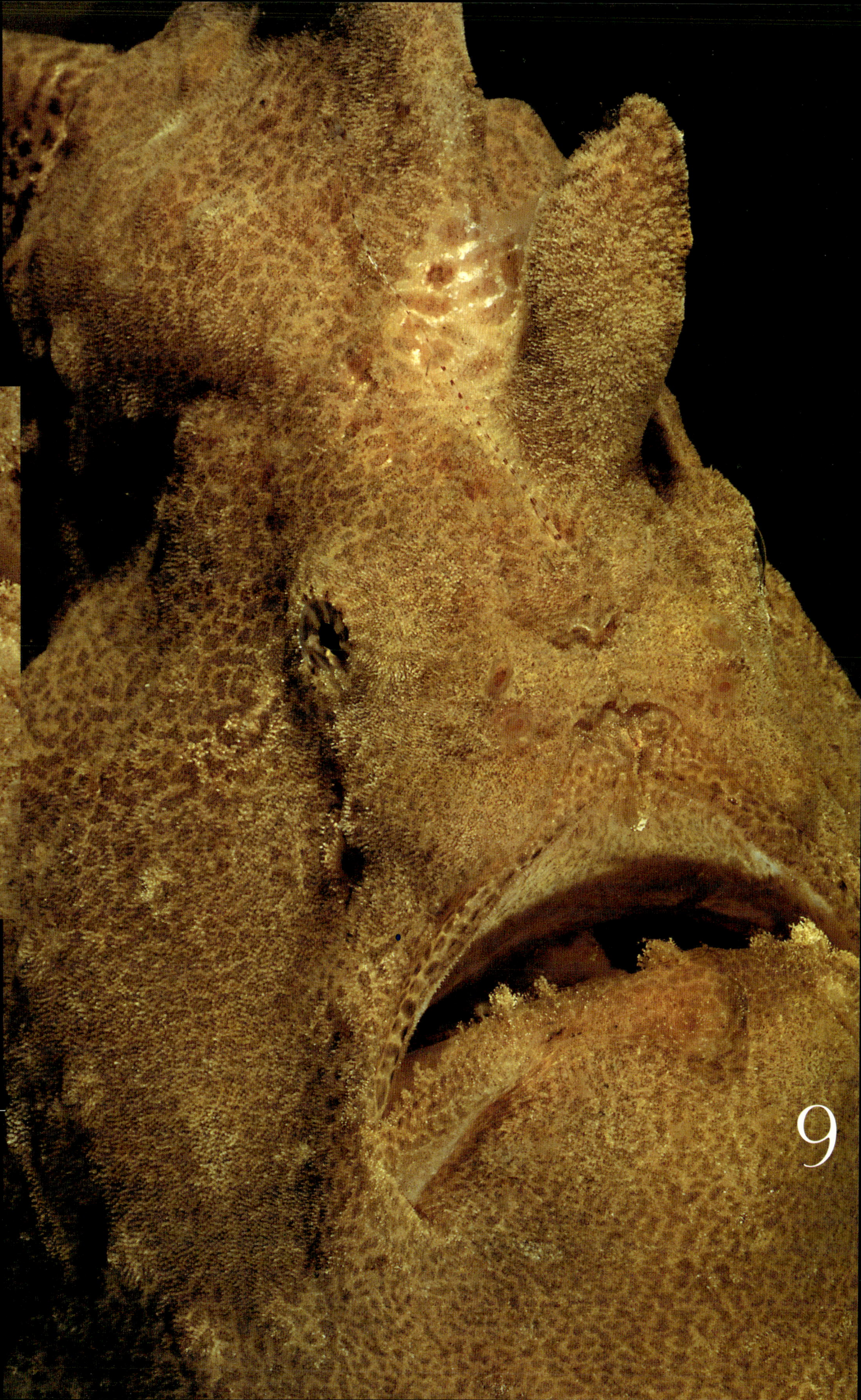

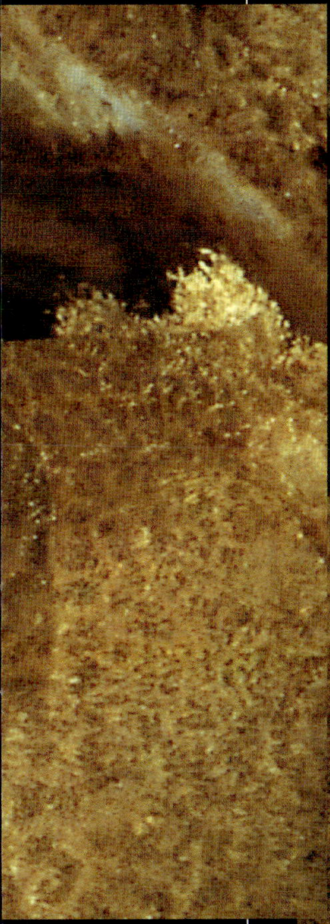

9

Blasenkorallen-Garnele
Vir philippinensis
Bubble coral shrimp
2 cm

12

Zwerg-Seepferdchen
Hippocampus sp.
Pigmy seahorse
1 cm

13

Krokodil-Schlangenaal
Branchysomophis crocodilinus
Crocodile snake eel
80 cm

Sterngucker-Schlangenaal
Brachysomophis cirrocheilos
Ghost snake eel
125 cm

16

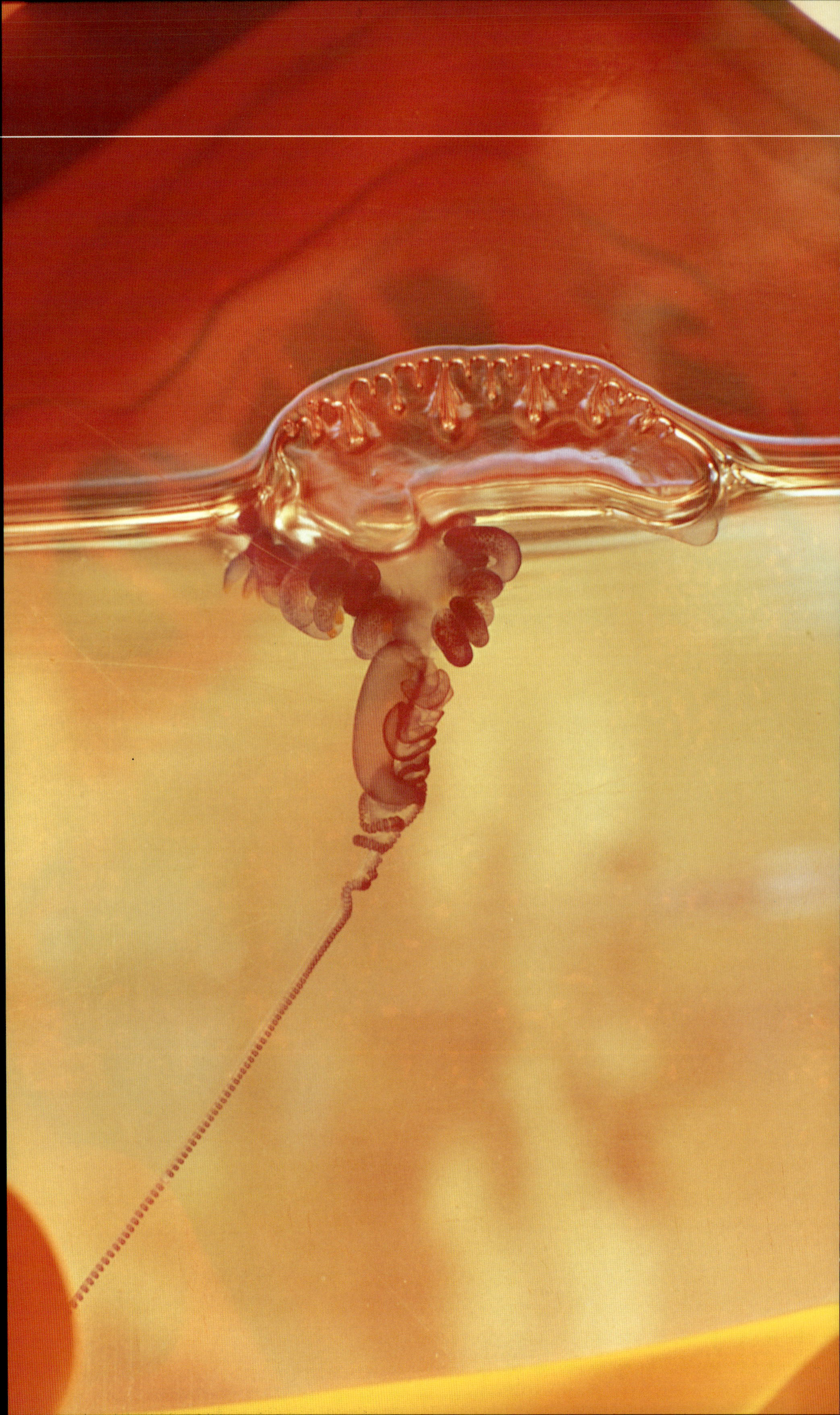

Portugiesische Galeere

Physalia physalis

Portugese man-of-war

3 cm

17

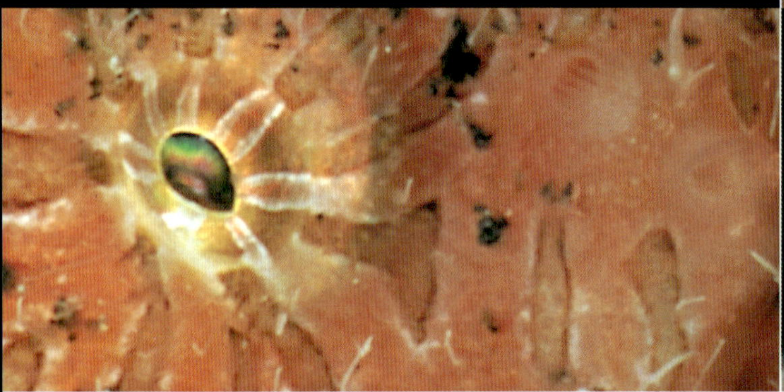

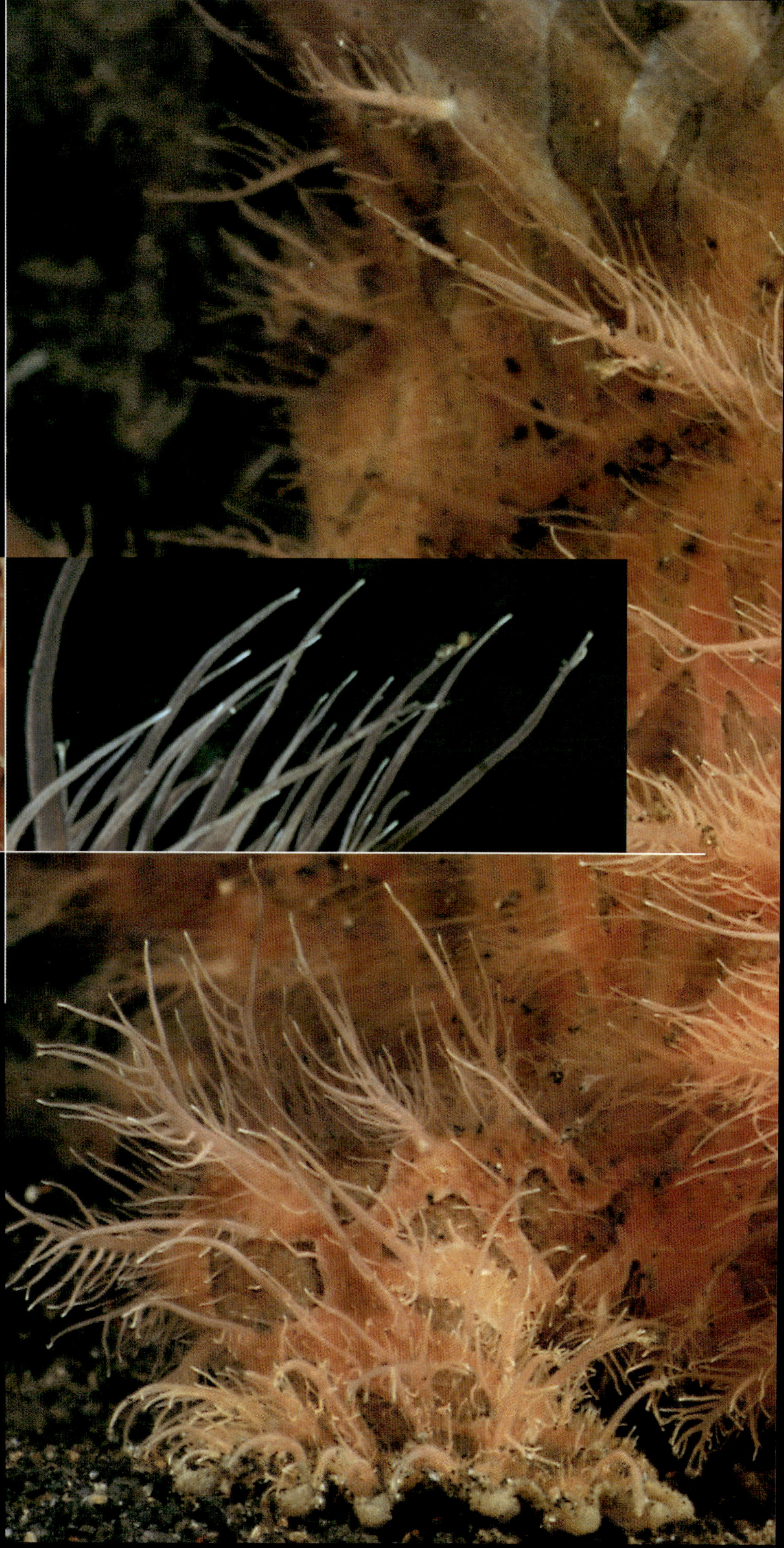

Gestreifter Froschfisch
Antennarius striatus
Striated frogfish
12 cm

Geistermuräne
Rhinomuraena quaesita
Ribbon eel
120 cm

Buckeldrachenkopf
Scorpaenopsis diabolus
Devil scorpionfish
20 cm

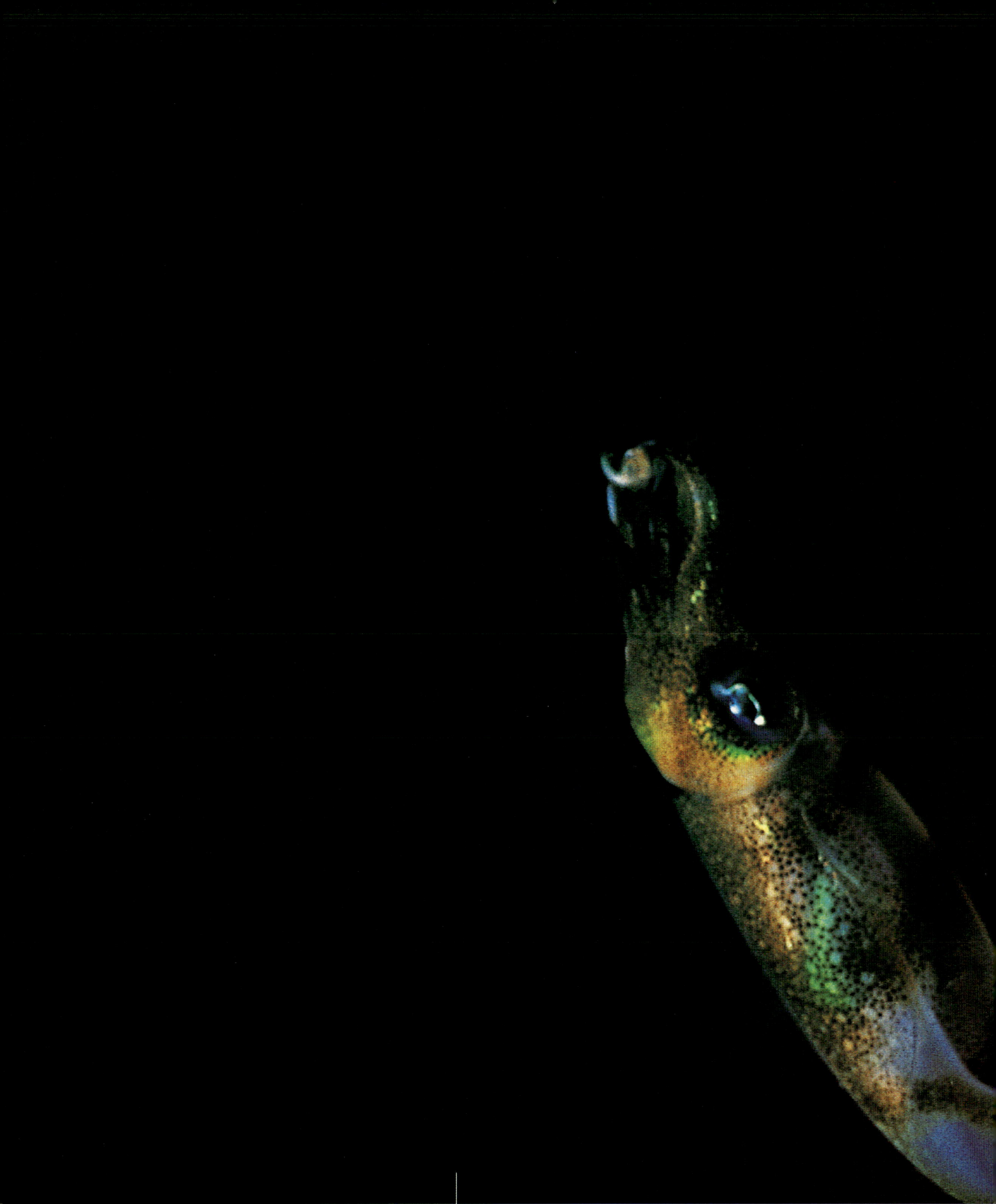

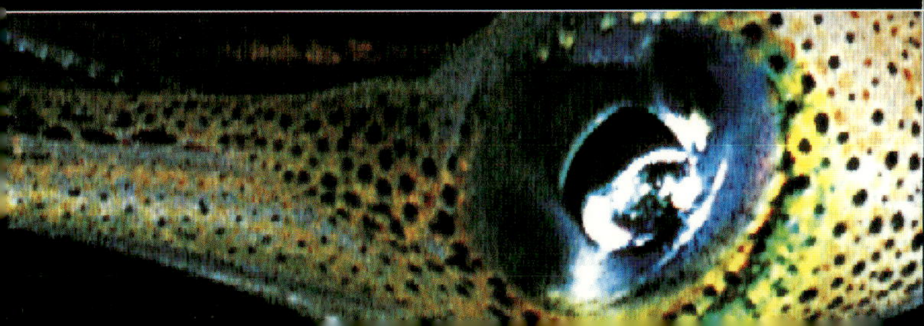

Großflossen-Kalmar
Sepiotheuthis lessoniana
Bigfin reef squid
23 cm

Blaugeringelter Octopus
Haplochlaena lunulata
Bluering octopus
10 cm

25

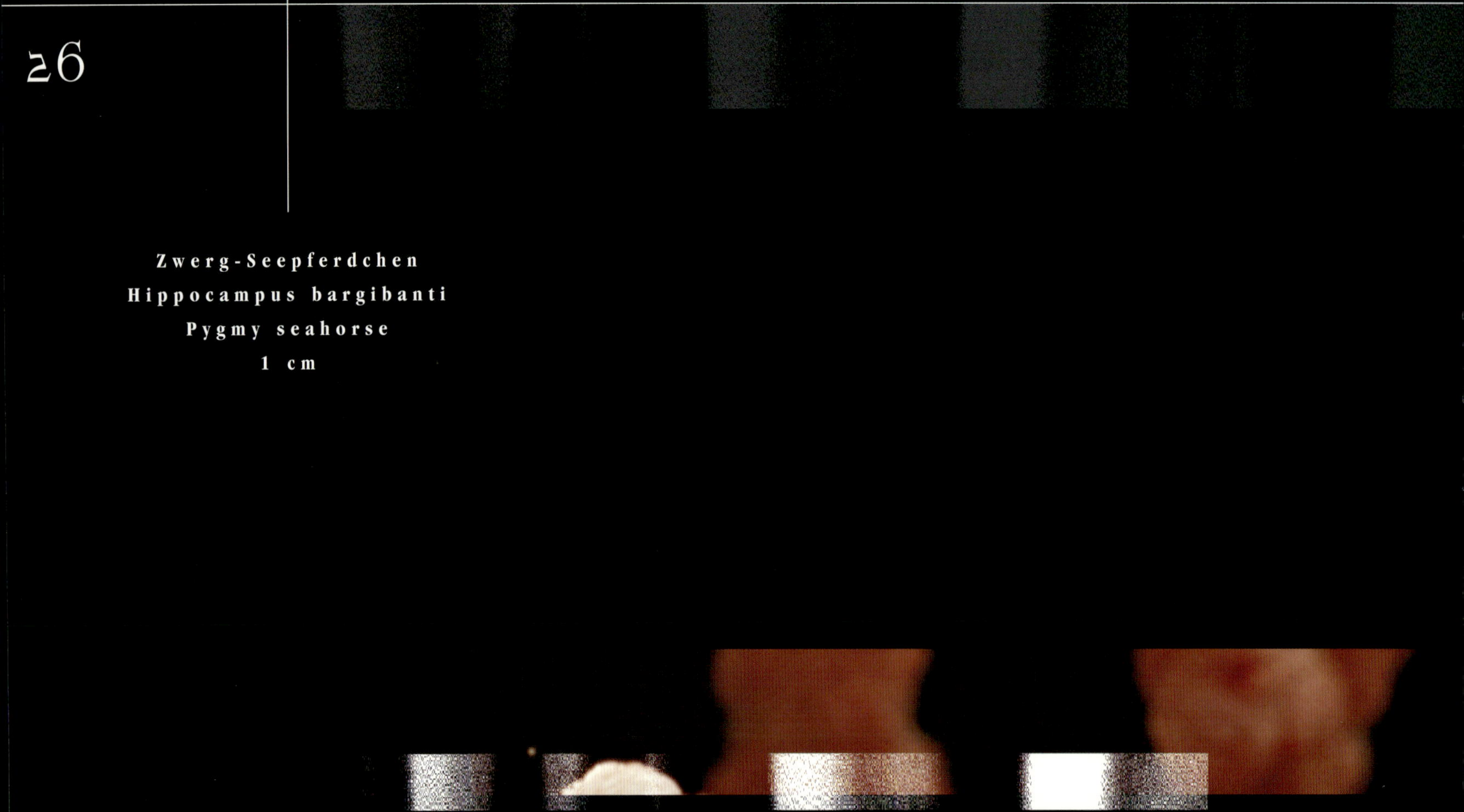

Zwerg-Seepferdchen
Hippocampus bargibanti
Pygmy seahorse
1 cm

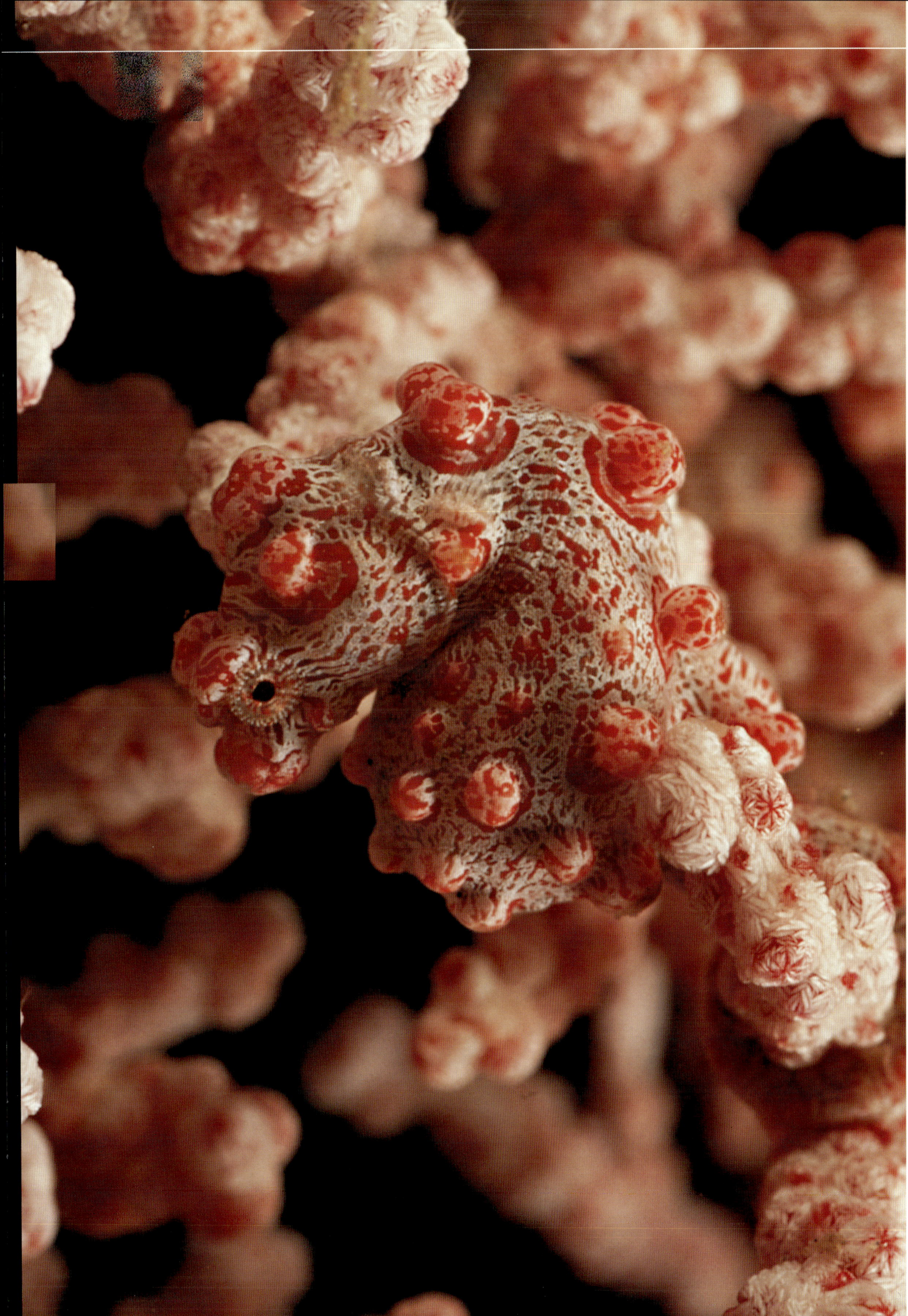

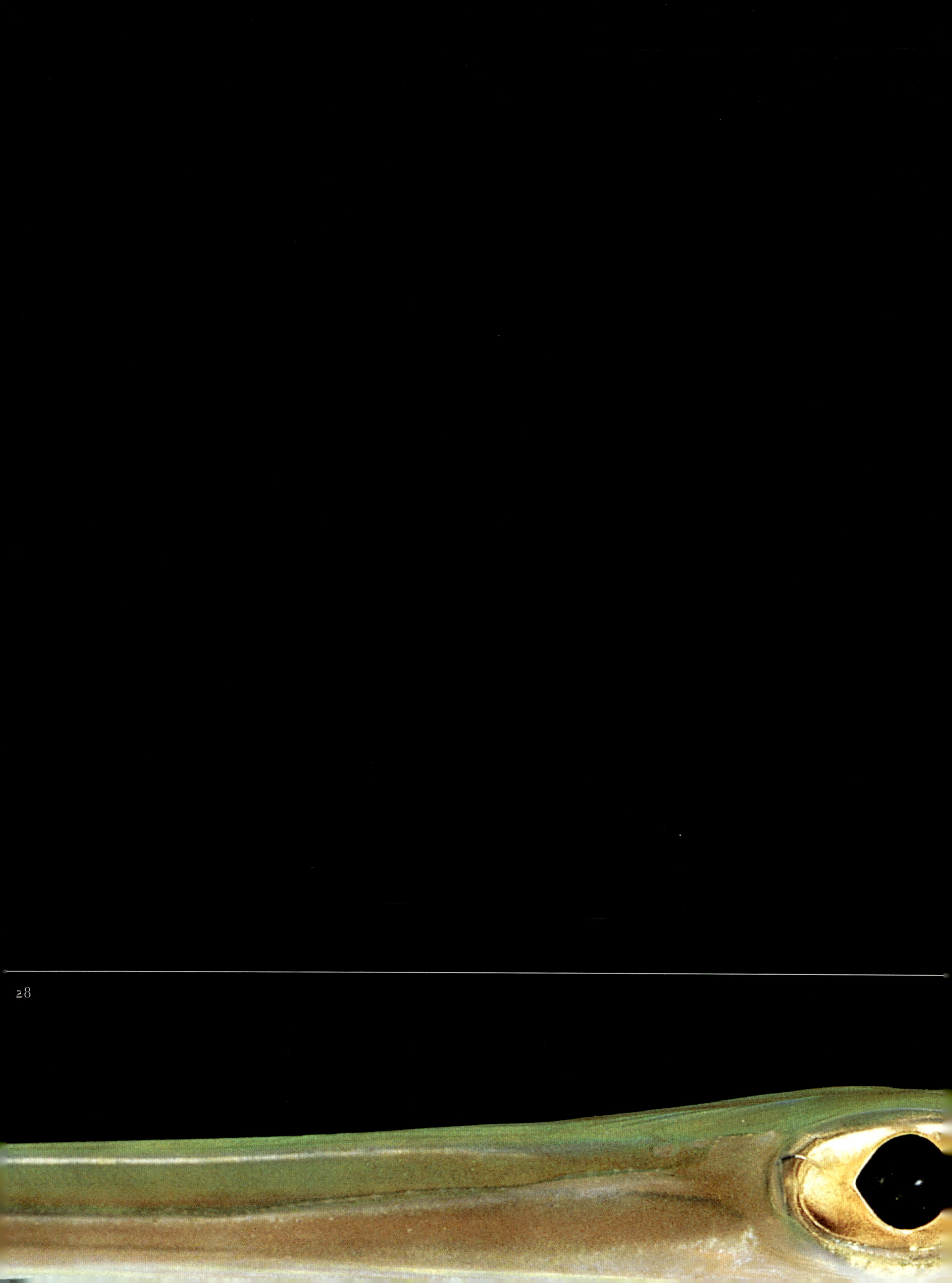

30

Gespensterkrebs
Caprella sp.
Skeleton shrimp
2 cm

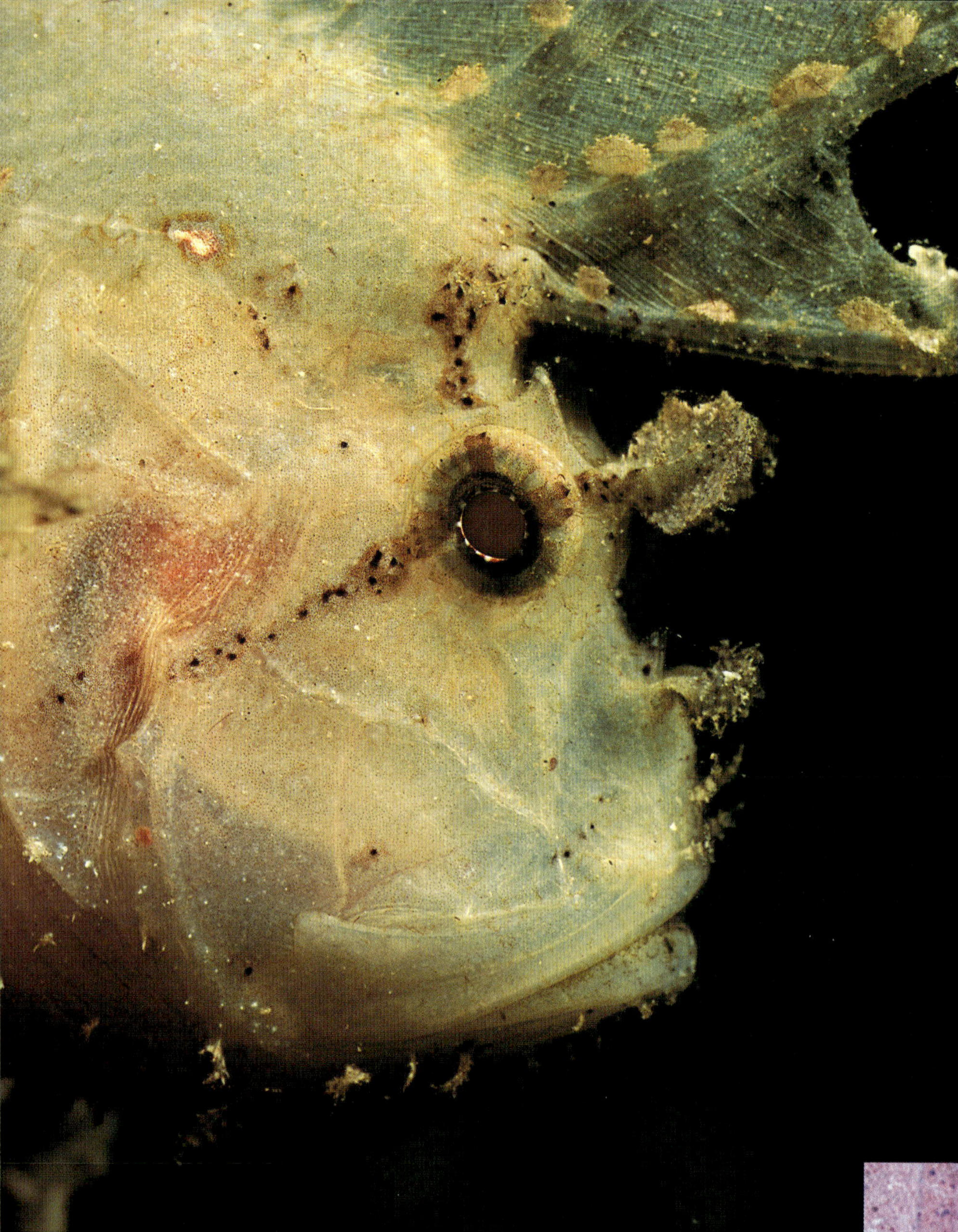

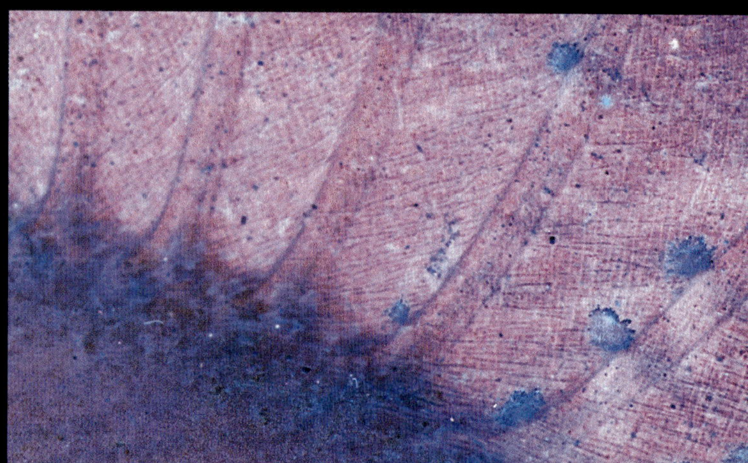

Großer Schaukelfisch

Taenianotus tricanthus

Leaf fish

Schwarzkorallen-Garnele
Tozeuma armatum
Black coral shrimp
4 cm

Igelfisch
Diodon sp.
Procupine fish
14 cm

Korallen-Seenadel
Corythoichthys intestinalis
Australian banded pipefish
10 cm

Schwarzfleck-Kugelfisch
Arothron nigropunctatus
Black-spotted pufferfish
22 cm

Porzellankrebs
Porcellanella picta
Seapen crab
2 cm

Partnergarnele
Hamodactylus sp.
Gorgonian shrimp
2 cm

Schiffshalter
Echeneis naucrates
Shark sucker
30 cm

46

Kleiner Teufelsfisch
Inimicus dicactylus
Pop-eyed sea goblin
20 cm

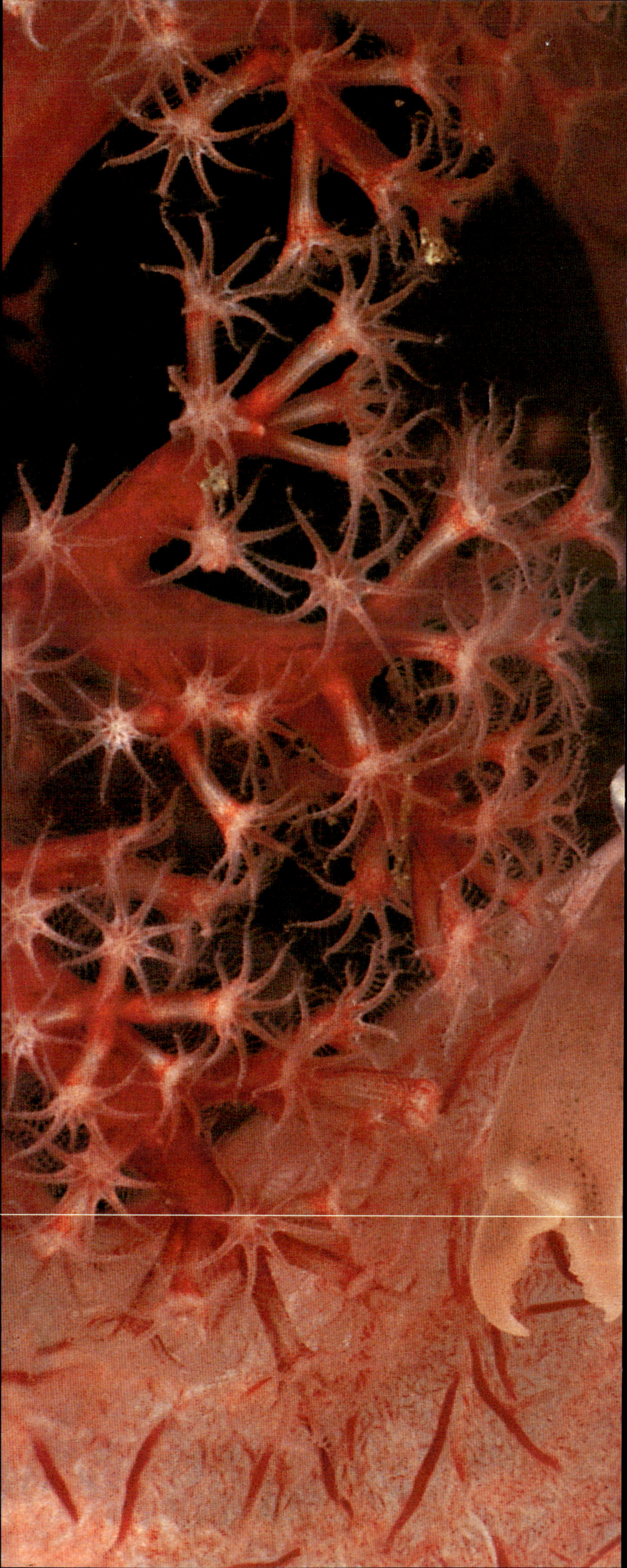

Porzellankrebs
Porcellanella picta
Seapen crab
2 cm

50

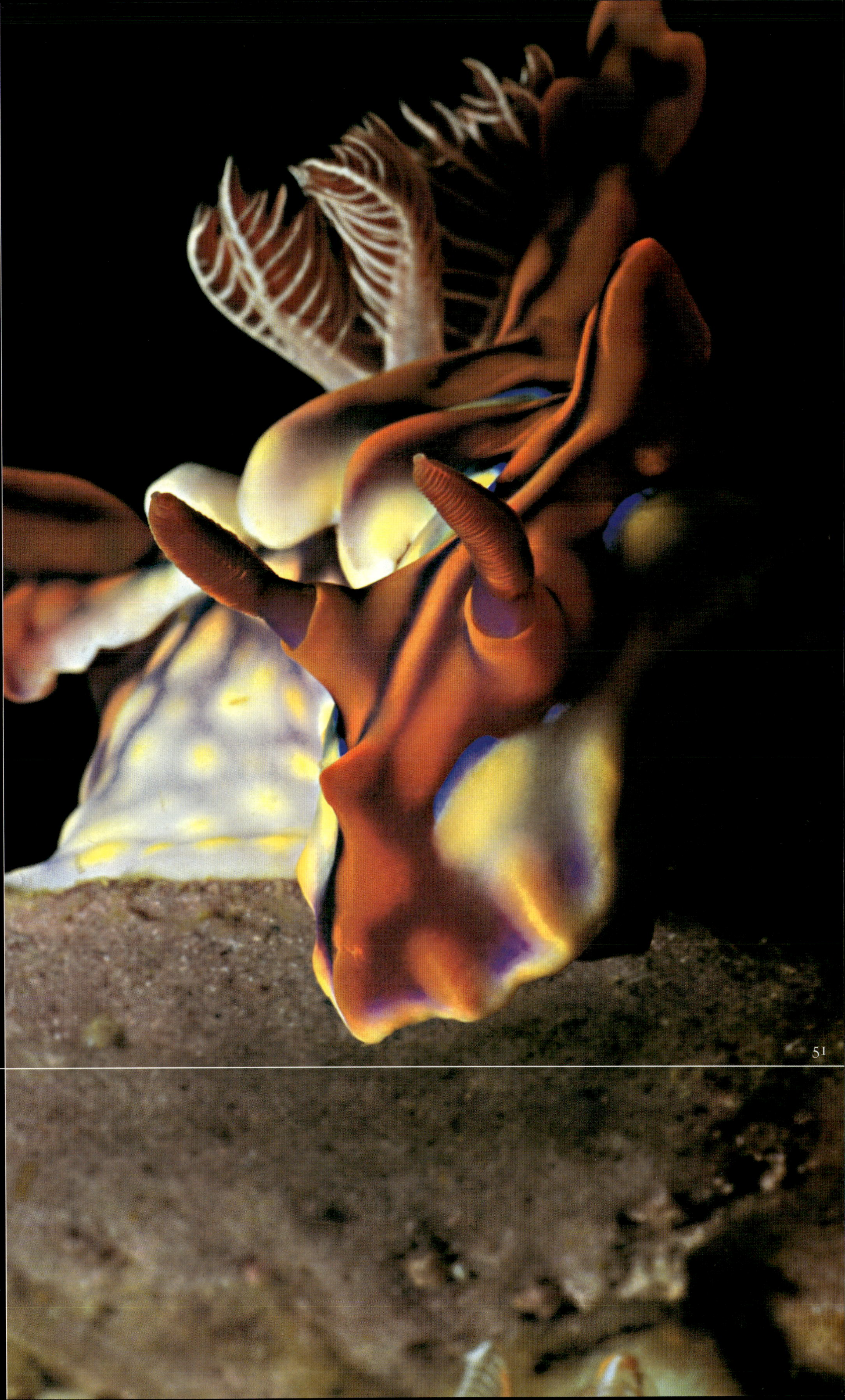

acht-Nacktschnecke
iamira magnifica
Nudibranch
4 cm

Gebänderter Fangschreckenkrebs
Lysiosquilla maculata
Tiger mantis
18 cm

55

Fischassel
Anilocra sp.
Parasitic isopod
2 cm

Tentakel-Skorpionfisch
Rhinopias frondosa
Weedy Scorpionfish
20 cm

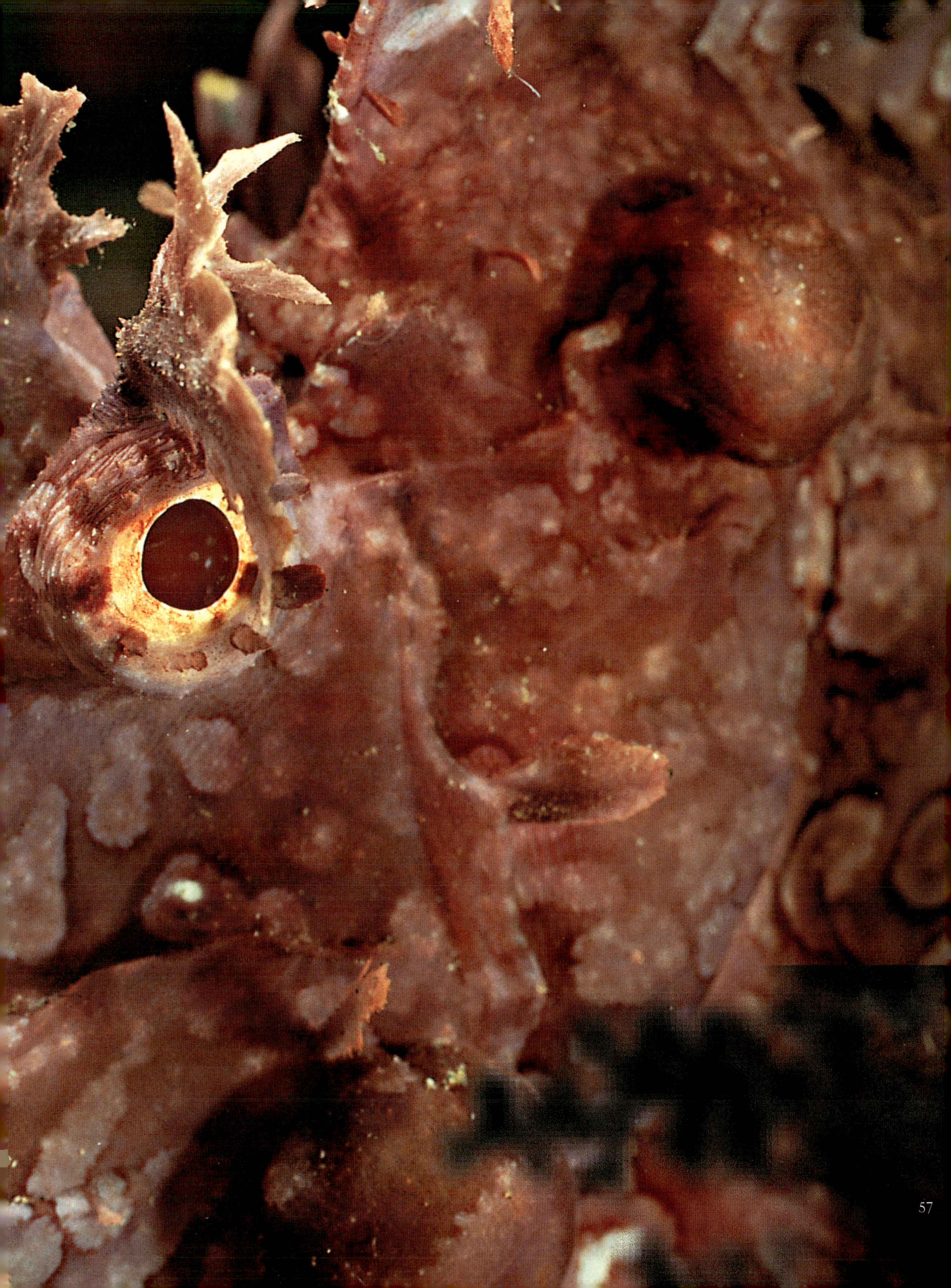

Warzige Schirmschnecke
Umbraculum umbraculum
Umbraculid
10 cm

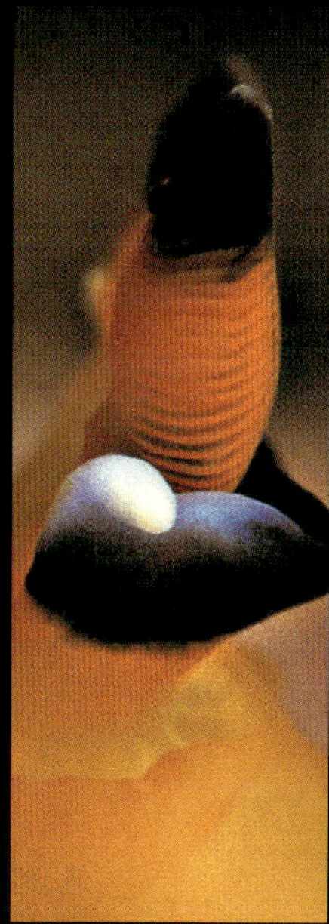

Pazifische Neonschnecke

Thecacera pacifica

Nudibranch

4 cm

61

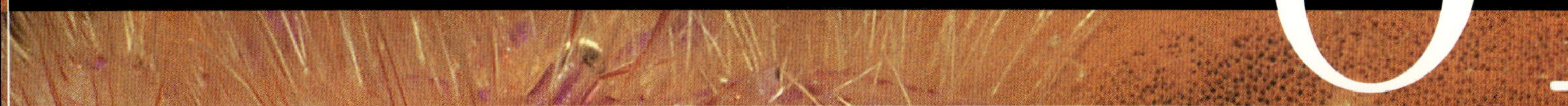

Violetter Springkrebs
Lauriea siagiani
Pink fairy lobster
2 cm

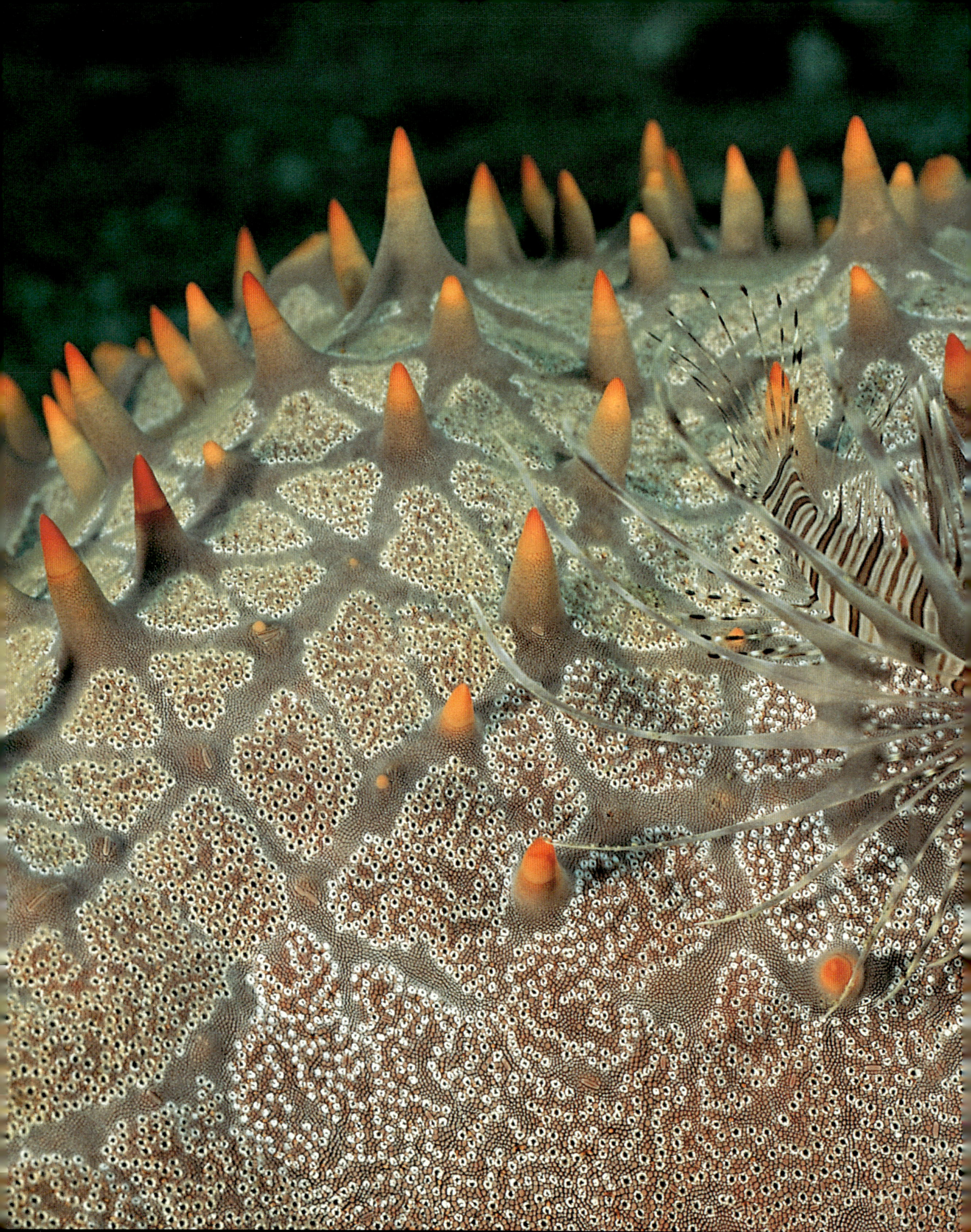

Rotfeuerfisch
Pterois volitans
Turkey fish
2 cm

64

Pilzkorallen-Seenadel
Siokunichthys nigrolineatus
Coral pipefish
8 cm

Dorniges Seepferdchen
Hippocampus hystrix
Spiny seahorse
10 cm

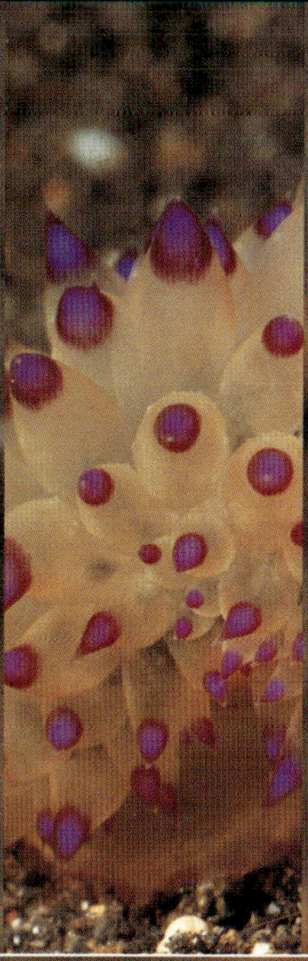

Prachtkolbenschnecke
Janolus sp.
Nudibranch
4 cm

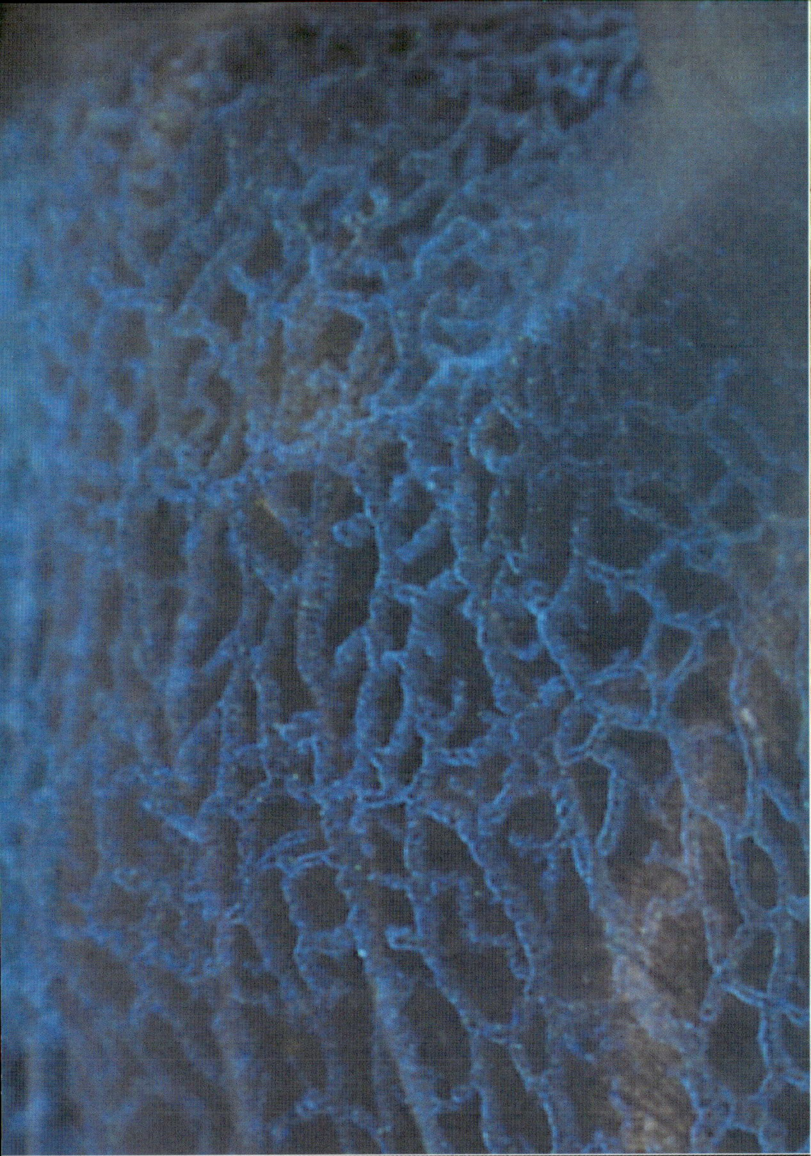

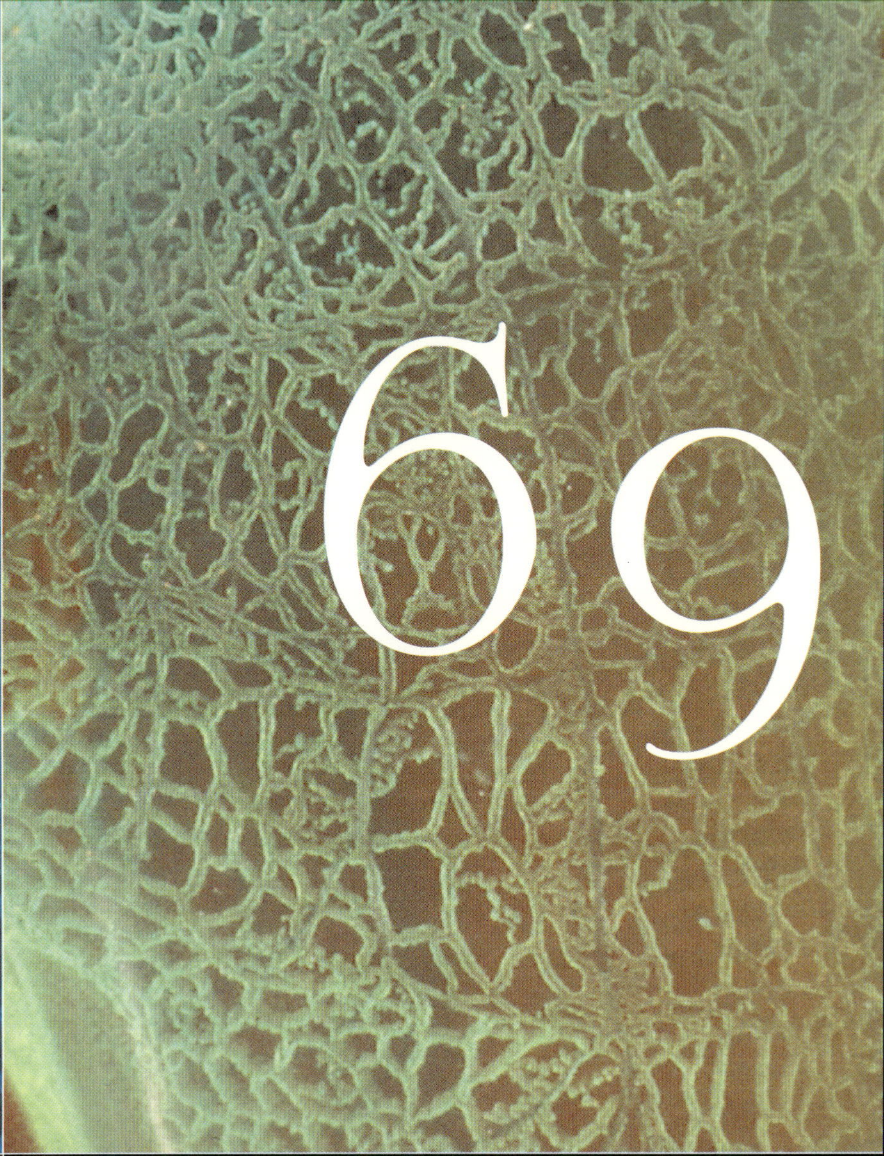

Solitäre Kannenseescheide
Rhopalaea sp.
Sea squirt
2 cm

Ambon Skorpionfisch
Pteroidichthys amboinensis
Ambon scorpionfish
8 cm

Korallen-Einsiedlerkrebs
Dardanus sp.
Hermit crab
3 cm

innenkrebs
Majidae
pider crab
2 cm

Peitschenkorallen-Zwerggrundel
Bryaninops youngei
Wire coral goby
3 cm

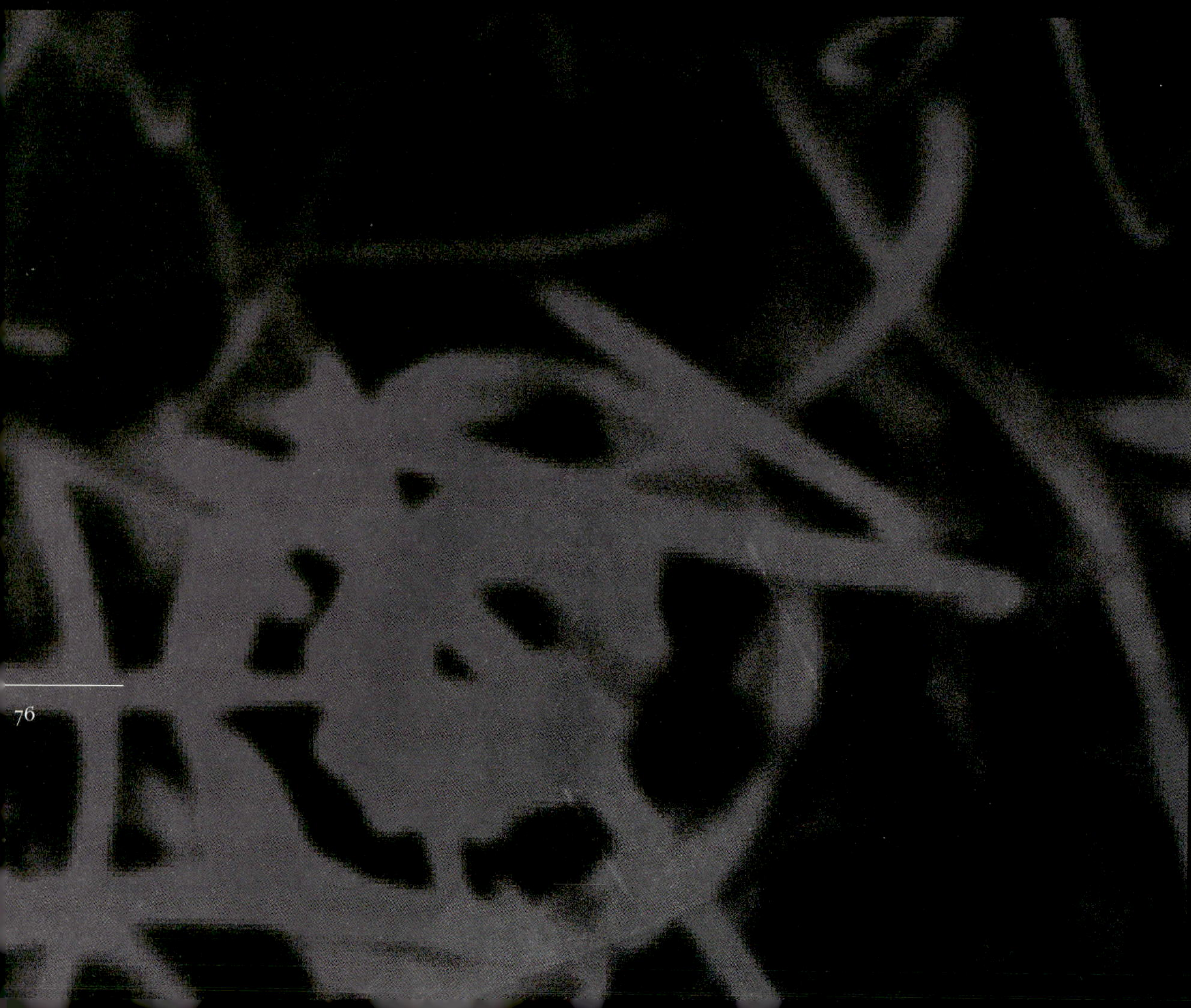

77

Große Tintenschnecke
Sepia sp.
Cuttlefish
35 cm

78

Hohlkreuz-Garnele

Thor amboinensis

Sexy shrimp

2 cm

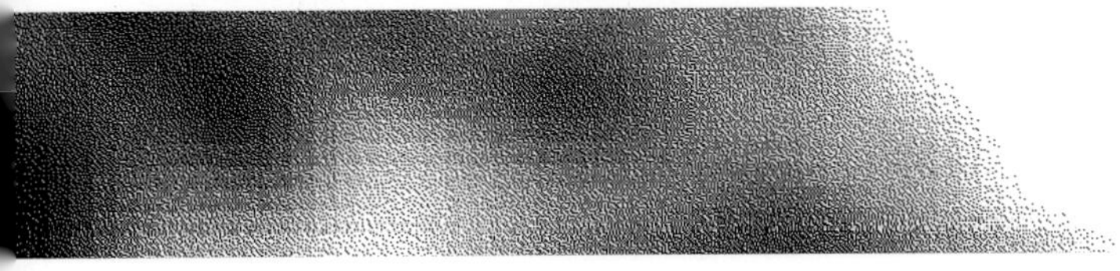
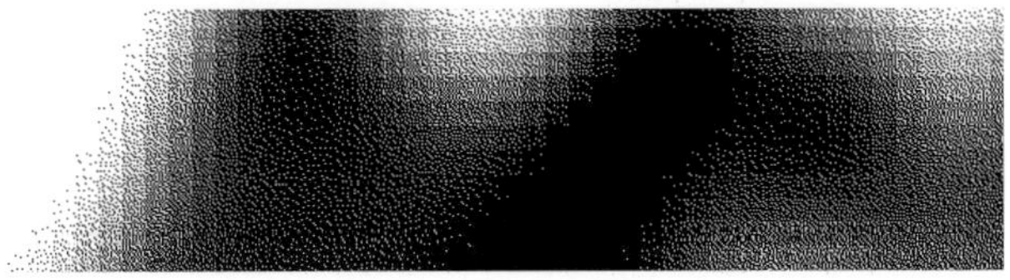

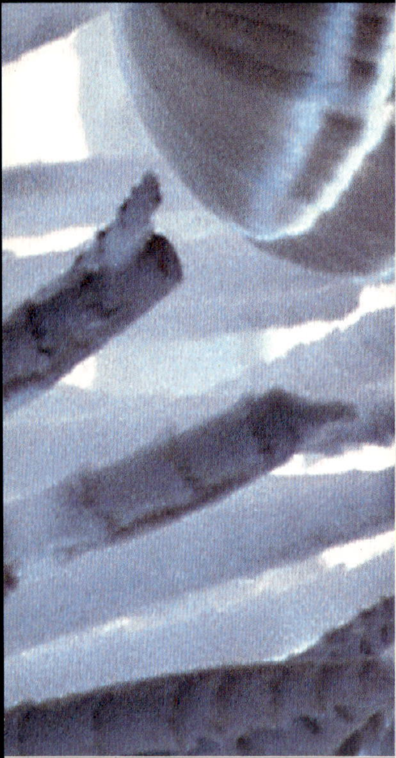

Federstern-Springkrebs
Allogalathea elegans
Elegant squad lobster
2 cm

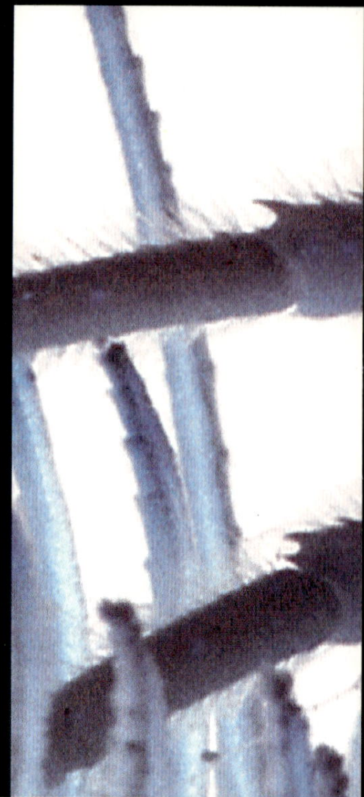

81

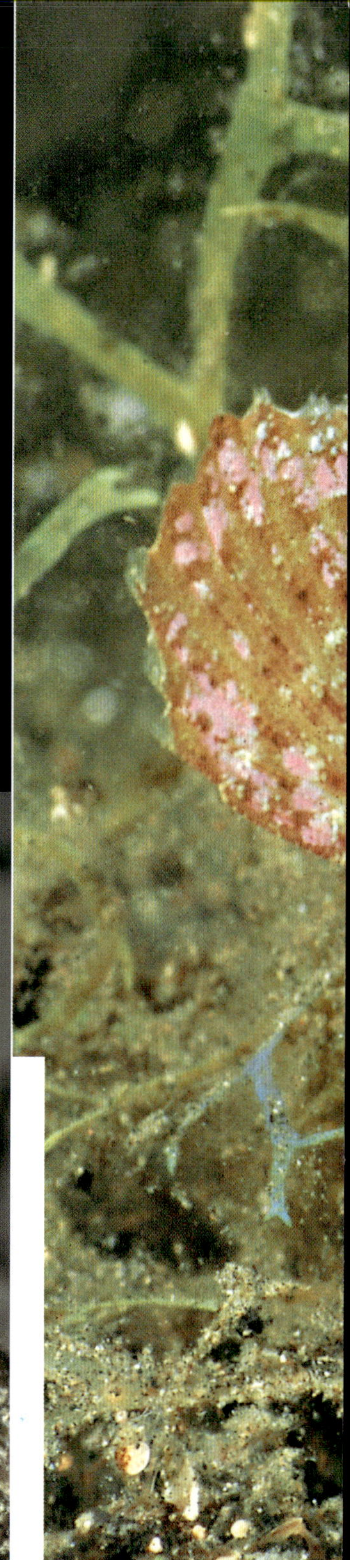

Seegras-Seenadel
Corythoichthys sp.
Seaweed pipefish
18 cm

Kakadu-Schaukelfis
Ablabys taenianotu
Cockatoo wasp fisl
7 cm

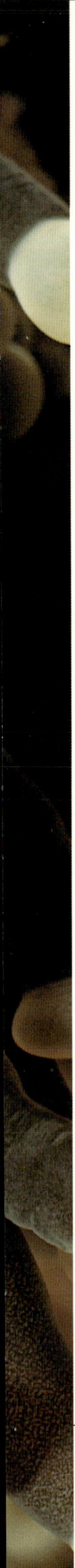

85

Japanischer Spinnenkrebs
Acheus japonicus
Organ utan crab
2 cm

Weißgesicht-Wespenfisch
Richardsonichthys leucogaster
White faced waspfish
12 cm

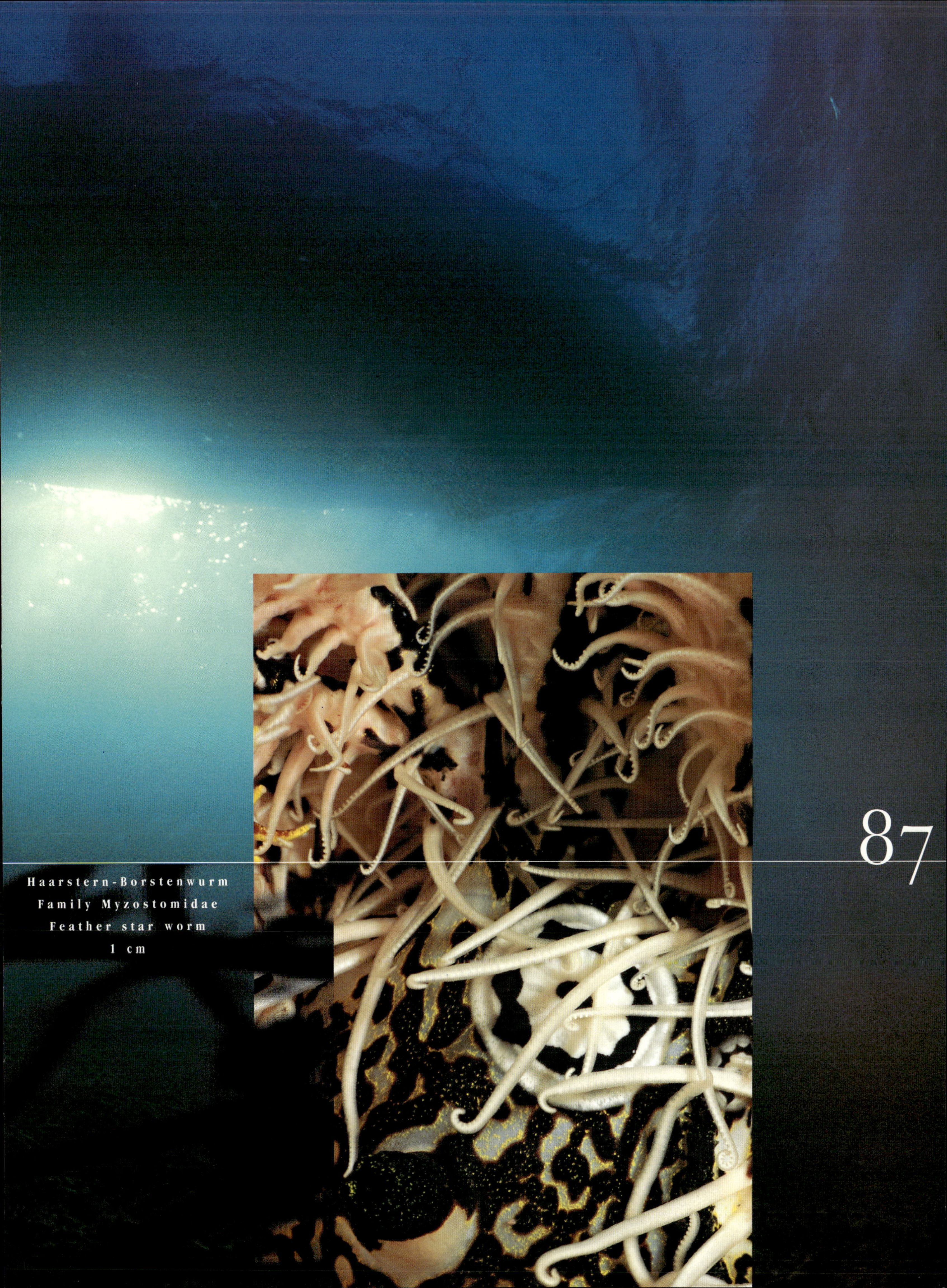

87

Haarstern-Borstenwurm
Family Myzostomidae
Feather star worm
1 cm

Krabbenaugen-Grundel
Signigobius biocellatus
Crab-eyed goby
8 cm

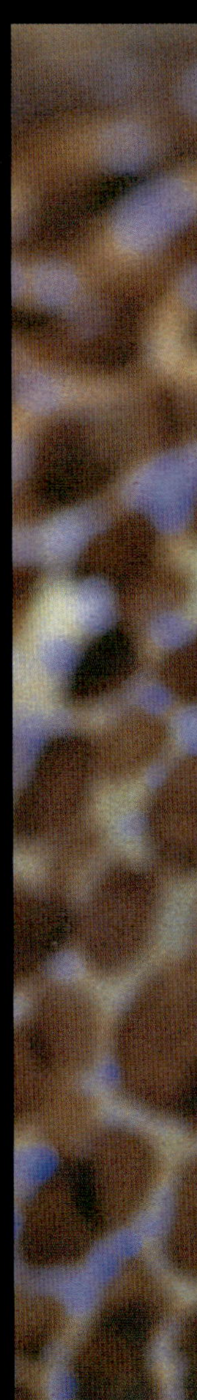

Kleine Tintenschnecke
Euprymna sp.
Bobtail squid
3 cm

Partner-Garnele
Allopontonia iaini
Urchin shrimp
2 cm

Zebra-Octopus
Octopus chierchiae
Mimic octopus
60 cm

Fetzen-Geisterfisch

Solenostomus paradoxus

Phantom ghost pipefish

10 cm

Riesen-Froschfisch
Antennarius commersoni
Commerson's frogfish
20 cm

Clown-Froschfisch
Antennarius maculatus
Clown frogfish
9 cm

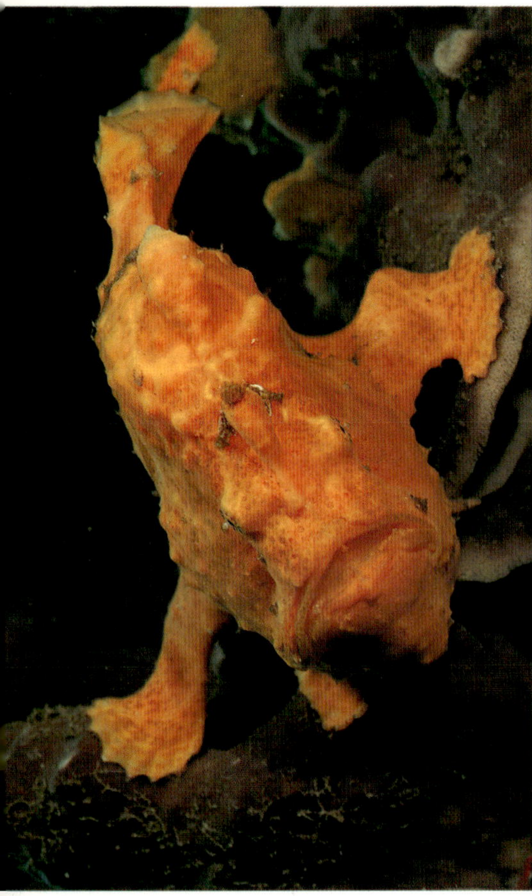

Clown-Froschfisch
Antennarius maculatus
Clown frogfish
20 cm

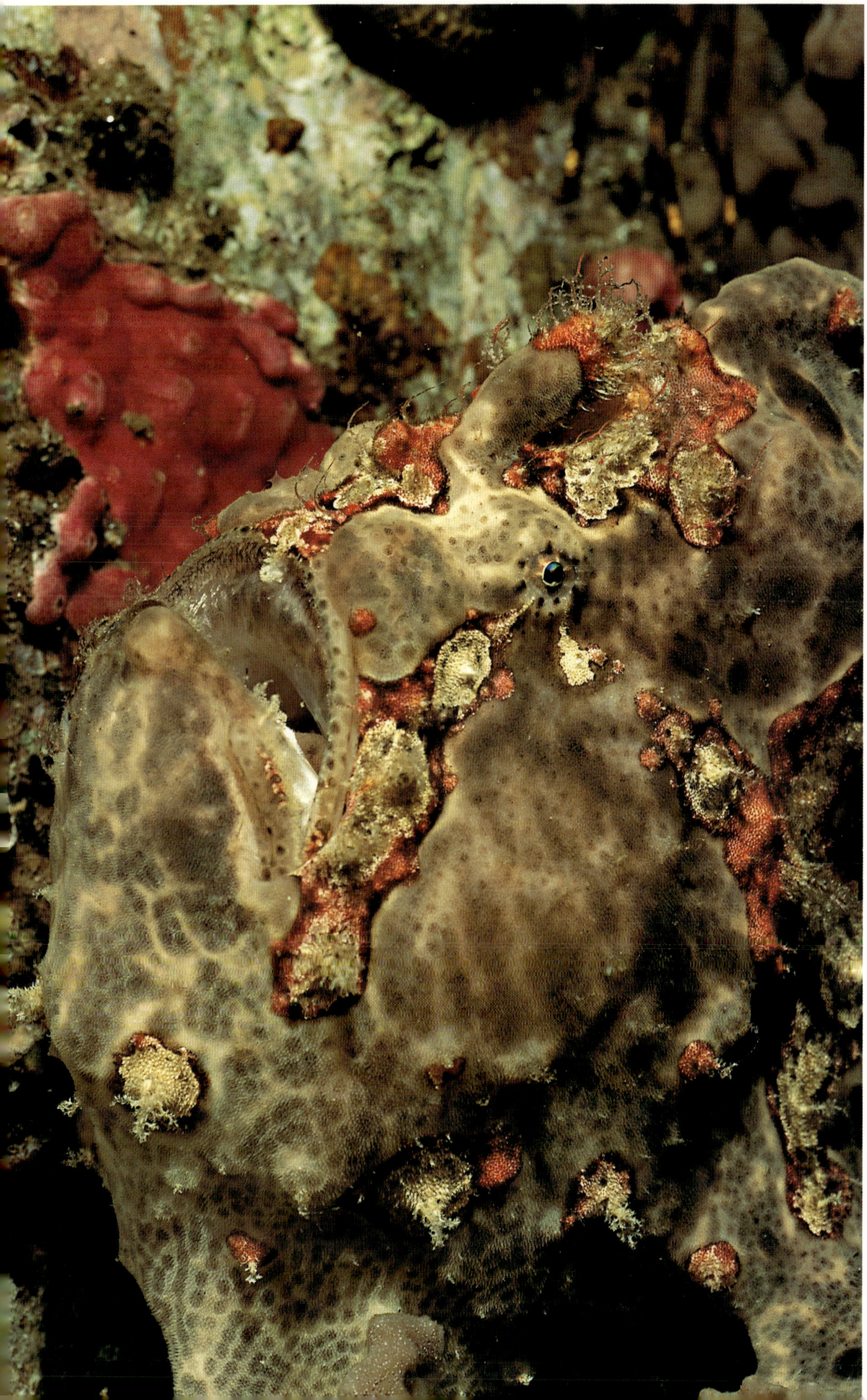

99

Pfeffers Tintenschnecke
Metasepia pfefferi
Flambuoyant cuttlefish
6 cm

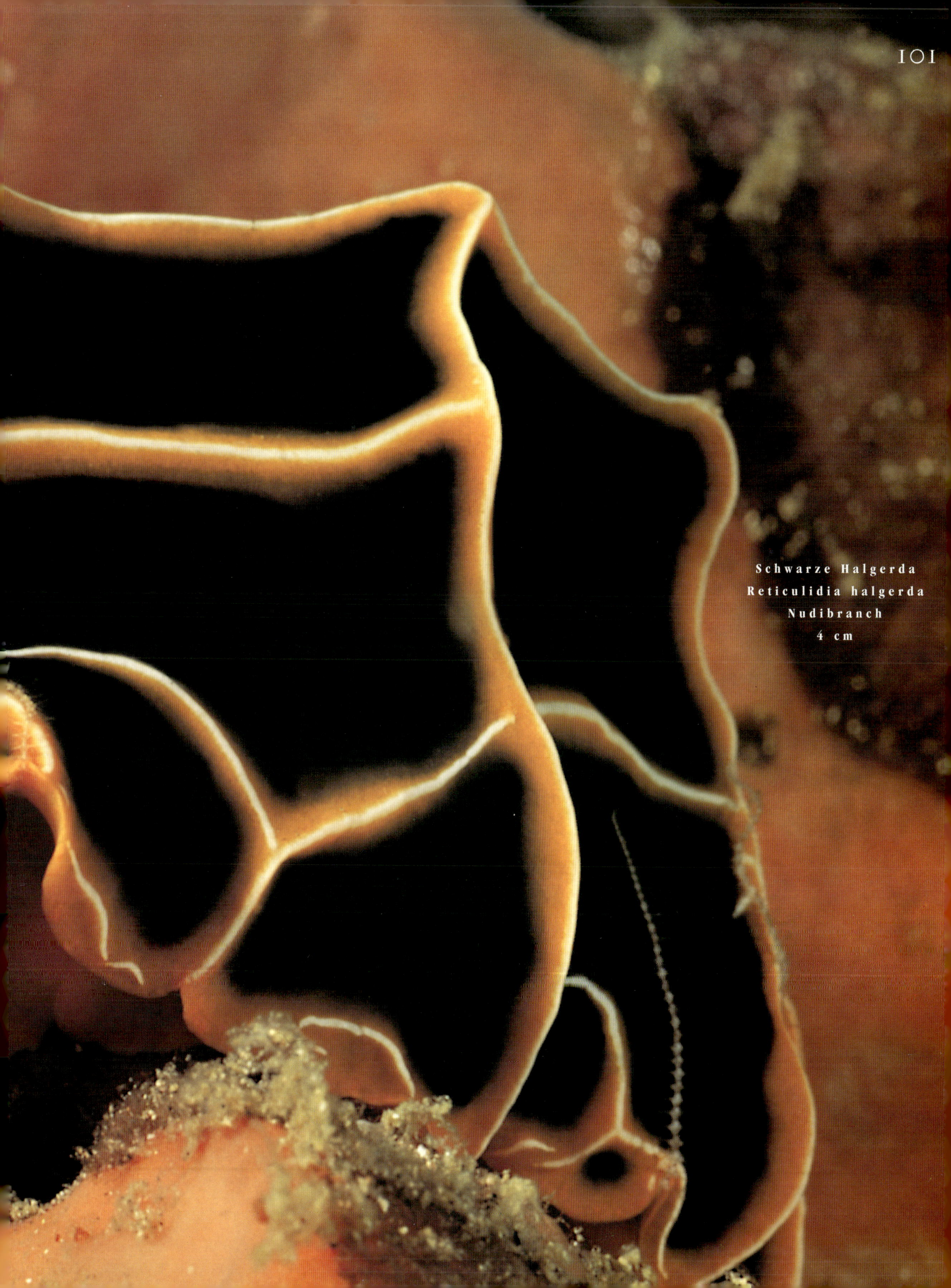

Schwarze Halgerda
Reticulidia halgerda
Nudibranch
4 cm

Kurzflossen-Zwergfeuerfisch
Dendrochirus brachyopterus
Fuzzy dwarf lionfish
12 cm

Krokodilfisch
Platycephalid fish
Crocodile fish, Flathead
20 cm

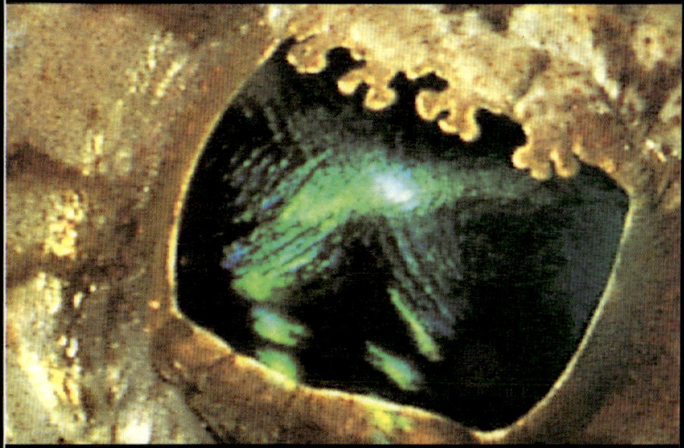

105

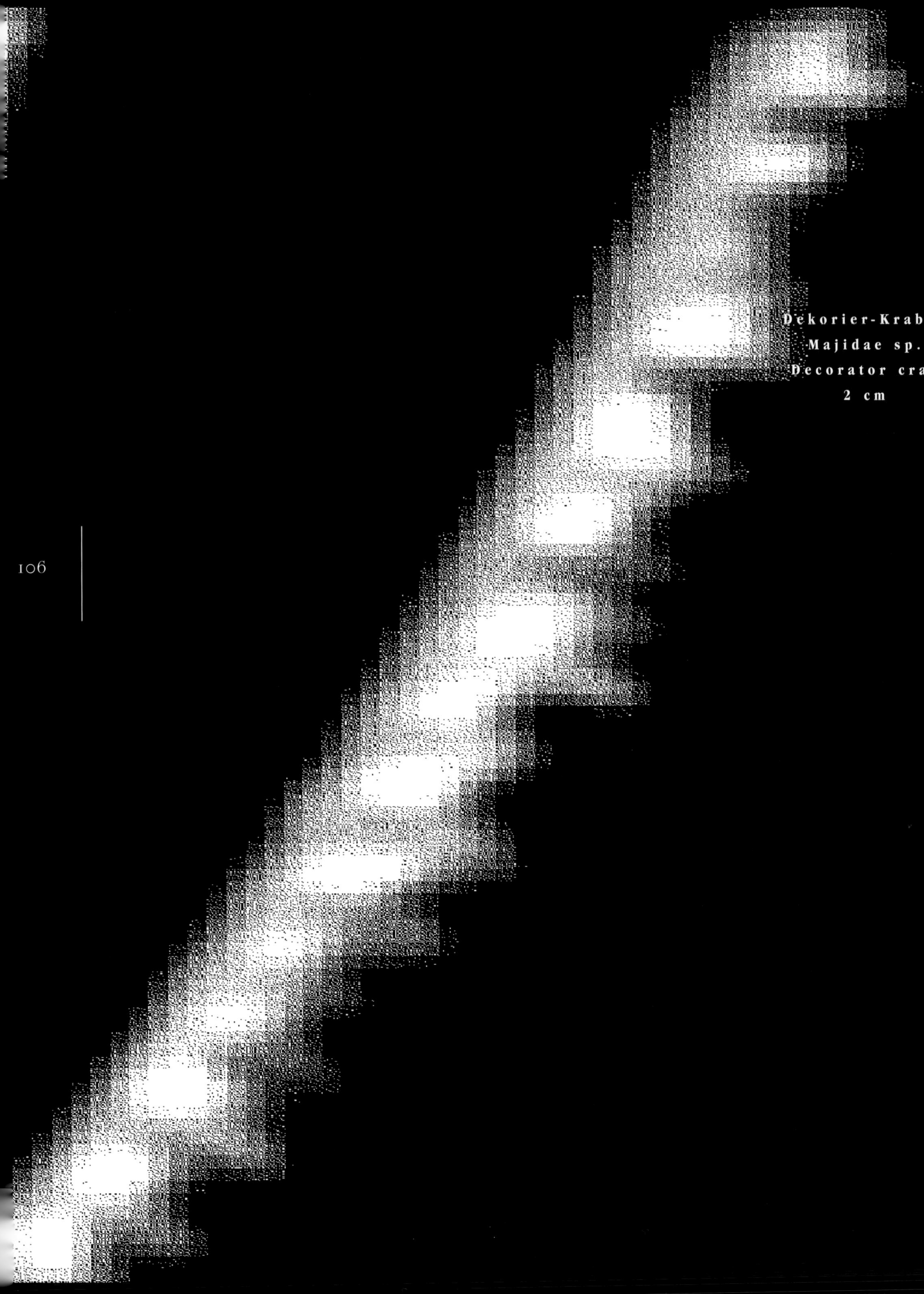

Dekorier-Krabbe
Majidae sp.
Decorator crab
2 cm

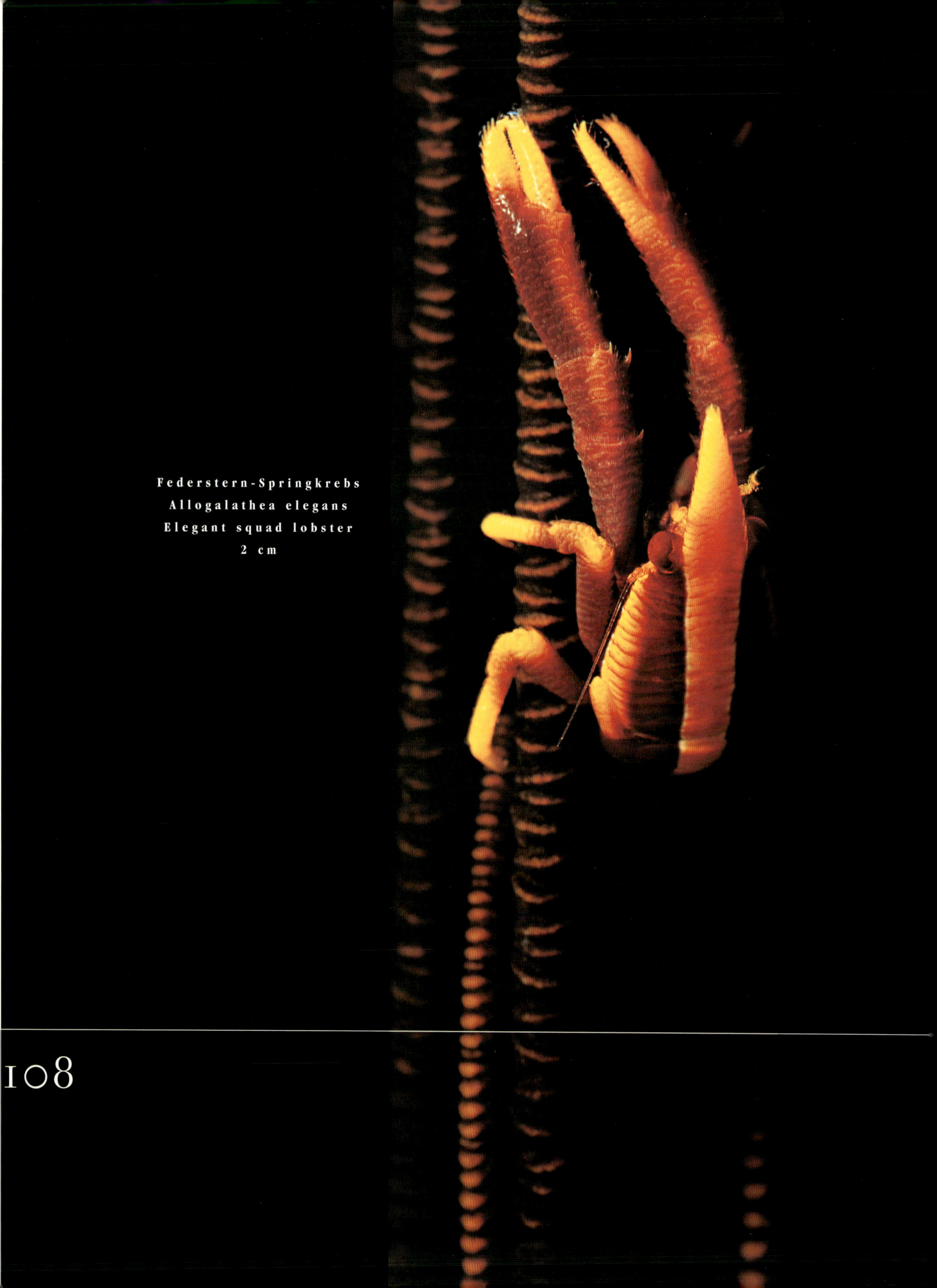

Federstern-Springkrebs
Allogalathea elegans
Elegant squad lobster
2 cm

Spinnenkrabbe
Xenocarcinus conicus
Spider crab
2 cm

Spinnenkrabbe
Majidae sp.
Spider crab
2 cm

Stachelige Spinnenkrabbe
Xenocarcinus tuberculatus
Black coral crab
2 cm

Drahtkorallen-Garnele
Dasycaris zanzibarica
Sea whip crab
1 cm

14

Spinnenkrabbe
Majidae sp.
Spider crab

Buckeldrachenkopf
Scorpaenopsis diabolus
Devil scorpionfish
20 cm

Marmor-Garnele
Saron neglectus
Marmorated shrimp
4 cm

Große Tintenschnecke
Sepia sp.
Cuttlefish
33 cm

120

Putzer-Garnel
Saron sp.
Cleaner shrim
4 cm

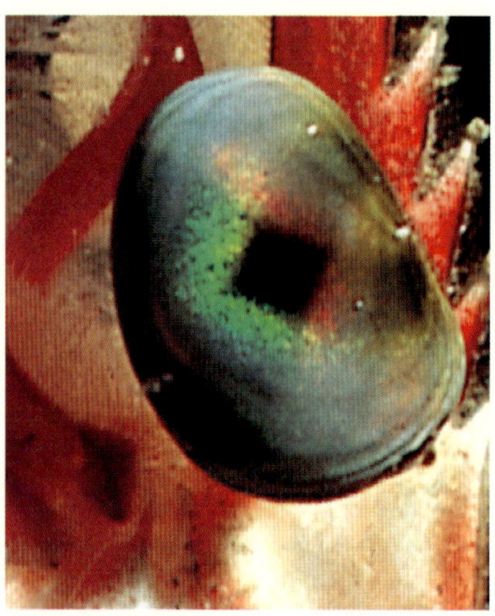

Blaugeringelter Octopus
Haplochlaena lunulata
Bluering octopus
10 cm

Seegras-Geisterfisch
Solenostomus cynaopterus
Robust ghost pipefish
9 cm

Große Hammerschnecke

Ceratosoma trilobatum

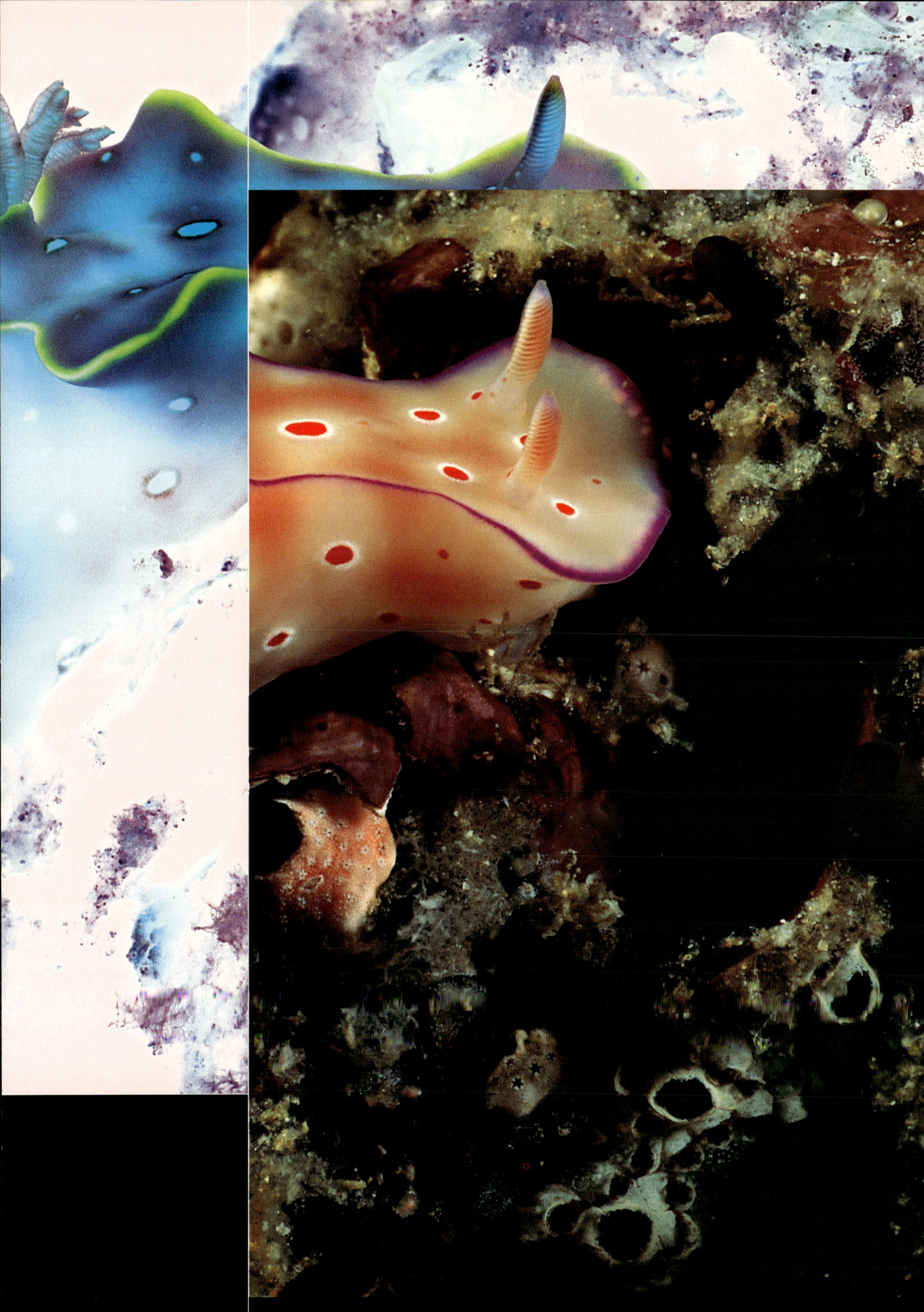

Gelblippen-Seekobra
Laticauda colubriana
Banded sea snake
120 cm

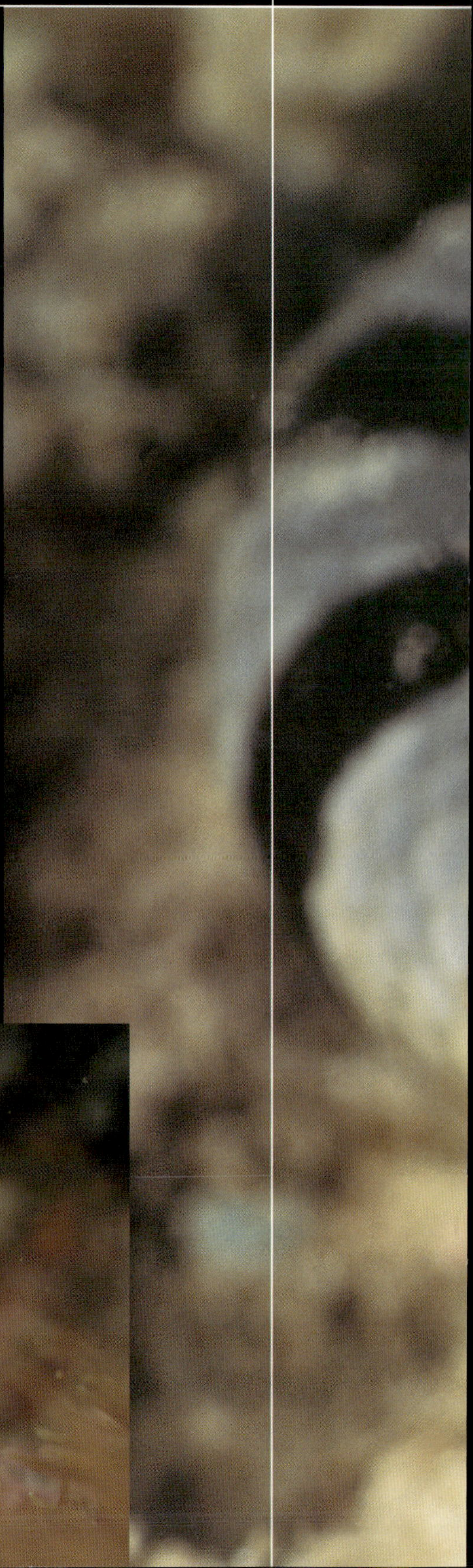

Eidechsenfisch
Synodus variegatus
Reef lizard fish
20 cm

Fadenschnecke
Moridilla brocki
Nudibranch
5 cm

130

Prachtkiemenschnecke
Miamira sp.
Nudibranch
4 cm

Putzer-Garnele
Saron sp.
Cleaner shrimp
4 cm

Tarnkrabbe
Cyclocoeloma tuberculata
Decorator crab
8 cm

Mondschnecke
Euselenops luniceps
Moonlight nudibranch
8 cm

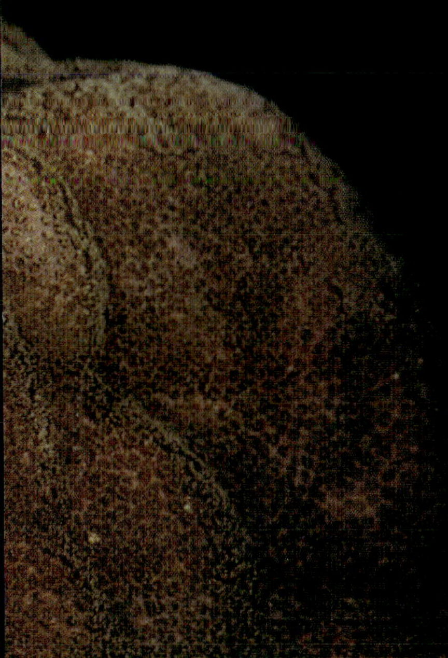

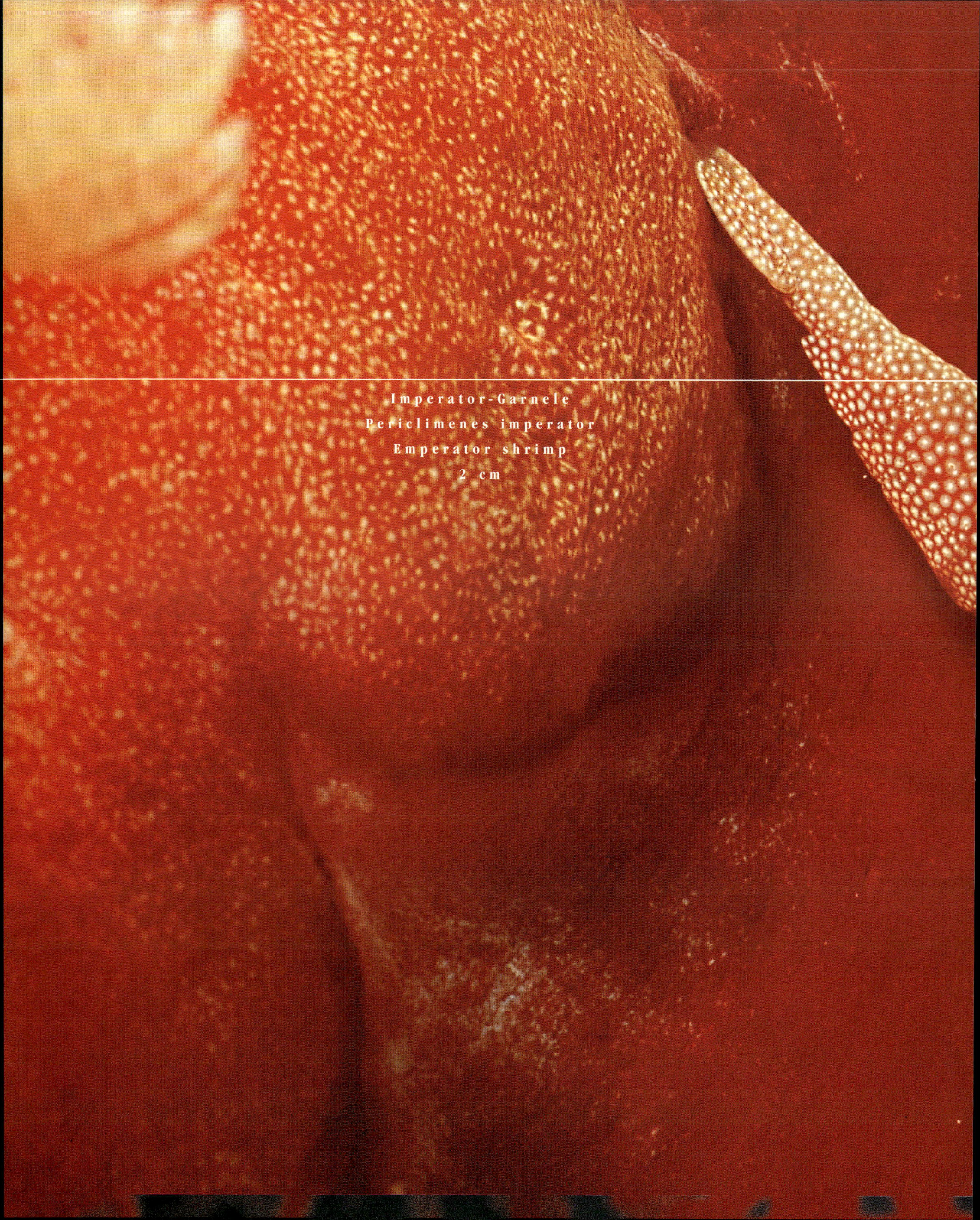

Imperator-Garnele
Periclimenes imperator
Emperator shrimp
2 cm

Feuerwurm
Chloeia sp.
Bristle worm
4 cm

Springkrebs
Galathea sp.
Squat lobster
2 cm

Blaue Schwimmkrabbe
Portunus pelagicus
Swimming crab
15 cm

140

141

Zebrakrebs
Zebrida adamsii
Zebra crab
2 cm

Flugelroßfisch

Eurypegasus dracon

Pegasus sea moth

8 cm

Herausgeber:
©2000
Monique Walker Dive Expeditions,
Hostatostraße 9, 65929 Frankfurt/Main, Deutschland
Tel.: +49 (0)69 / 30 41 51 Fax: +49 (0)69 / 30 09 24 60
info@dive-expeditions.com www.dive-expeditions.com

Idee, Fotografie und Text:
Monique Walker, Frankfurt/Main, Deutschland

Biologische Bestimmungen Deutsch/Latein:
Dr. Harry Erhardt, Fritzlar, Deutschland

Bearbeitung der Übersetzungen:
Max Gibbs, Oxford, GB

Gestaltung:
Bauer & Kowallik Communication GmbH, Frankfurt/Main, Deutschland

Lithografie:
Litho-Studio Frankfurt GmbH, Frankfurt/Main, Deutschland

Druck:
Sebald Sachsendruck, Plauen, Deutschland

Die Deutsche Bibliothek – CIP – Einheitsaufnahme
Walker, Monique:
Critters : mysterious twilight zone / Monique Walker. - Frankfurt am Main :
Dive Expeditions, 2000
ISBN 3-9807064-0-0

Publisher:
©2000
Monique Walker Dive Expeditions,
Hostatostraße 9, 65929 Frankfurt/Main, Germany
Tel.: +49 (0)69 / 30 41 51 Fax: +49 (0)69 / 30 09 24 60
info@dive-expeditions.com www.dive-expeditions.com

Idea, photos and text:
Monique Walker, Frankfurt/Main, Germany

Biological classification German/Latin:
Dr. Harry Erhardt, Fritzlar, Germany

Correction of translation:
Max Gibbs, Oxford, GB

Design:
Bauer & Kowallik Communication GmbH, Frankfurt/Main, Germany

Reproduction:
Litho-Studio Frankfurt GmbH, Frankfurt/Main, Germany

Print:
Sebald Sachsendruck, Plauen, Germany

Made in Germany
Critters – mysterious twilight-zone
ISBN 3-9807064-0-0